Routledge Revivals

Financing the New Federalism

Financing the New Federalism is the fifth in a series on the governance of metropolitan areas which aimed to improve the political organisation of metropolitan regions in America. Originally published in 1975, this particular study focusses on federal revenue sharing exploring its effects and implications with the purpose of providing a breadth of views on the subject for policy-makers. This title will be of interest to students of environmental studies.

I0028200

Financing the New Federalism

Revenue Sharing, Conditional Grants, and Taxation

Robert P. Inman, Martin McGuire, Wallace E. Oates, Jeffrey L. Pressman and Robert D. Reischauer

RFF PRESS
RESOURCES FOR THE FUTURE

First published in 1975
by Resources for the Future, Inc.

This edition first published in 2016 by Routledge
2 Park Square, Milton Park, Abingdon, Oxon, OX14 4RN
and by Routledge
711 Third Avenue, New York, NY 10017

Routledge is an imprint of the Taylor & Francis Group, an informa business

© 1975, Resources for the Future, Inc.

Publisher's Note
The publisher has gone to great lengths to ensure the quality of this reprint but points out that some imperfections in the original copies may be apparent.

Disclaimer
The publisher has made every effort to trace copyright holders and welcomes correspondence from those they have been unable to contact.

A Library of Congress record exists under LC control number: 75019367

ISBN 13: 978-1-138-12213-0 (hbk)
ISBN 13: 978-1-315-65060-9 (ebk)
ISBN 13: 978-1-138-12216-1 (pbk)

Financing the New Federalism

NO. 5 IN A SERIES ON

The Governance of Metropolitan Regions

WALLACE E. OATES, EDITOR

Financing the New Federalism

REVENUE SHARING, CONDITIONAL GRANTS, AND TAXATION

Papers by

ROBERT P. INMAN

MARTIN McGUIRE

WALLACE E. OATES

JEFFREY L. PRESSMAN

ROBERT D. REISCHAUER

Published for Resources for the Future, Inc.
By The Johns Hopkins University Press, Baltimore and London

Resources for the Future is a nonprofit corporation for research and education in the development, conservation, and use of natural resources and the improvement of the quality of the environment. It was established in 1952 with the cooperation of the Ford Foundation. Part of the work of Resources for the Future is carried out by its resident staff; part is supported by grants to universities and other nonprofit organizations. Unless otherwise stated, interpretations and conclusions in RFF publications are those of the authors; the organization takes responsibility for the selection of significant subjects for study, the competence of the researchers, and their freedom of inquiry.

This study is the fifth in a series of books from RFF dealing with the problems of metropolitan governance. The papers in it were presented at a meeting of the Metropolitan Governance Research Committee, sponsored by the Academy for Contemporary Problems, Columbus, Ohio. The secretary of the committee was John E. Jackson, Department of Government, Harvard University, and the organizer of the meeting and editor of the volume is Wallace E. Oates of the Department of Economics, Princeton University. The charts for this volume were drawn by Frank and Clare Ford. The manuscript was edited by Jo Hinkel.

RFF editors: Herbert C. Morton, Joan R. Tron, Ruth B. Haas, Jo Hinkel

Contents

Tables

Figures

Commission on the Governance of Metropolitan Regions*

Foreword

The publication of this fifth number in the series on The Governance of Metropolitan Regions marks RFF's continued commitment to the study of possibilities for improving the political organization of those dominant features of the American scene. The original effort was launched in the spring of 1970, while the papers in this volume were read at a meeting held in Columbus, Ohio, in the spring of 1974. Here the specific subject is federal revenue sharing, its effects and implications, while past volumes have dealt with government reform, minority perspectives, and public services. Our general aim, as it has been in the earlier numbers, is to provide breadth of view for policy makers and to stimulate other researchers to pursue specific topics in greater depth. In this case, however, the papers have a more immediate relevance because of the congressional debate on the future of revenue sharing, which is anticipated in the summer of 1975.

In all research enterprises, the final product, while it may bear the names of only one or two men, reflects contributions by many. No less so here, and I would be remiss if I did not express RFF's thanks to these valuable participants. In particular, we are grateful to the Academy for Contemporary Problems, which funded the largest share of the expenses connected with these papers and the related meeting of the Metropolitan Government Research Committee. The members of the committee itself have, of course, contributed mightily to the product we present here—especially by their critical comments, but also by their general support. Joseph L. Fisher, the committee chairman, and Lowdon Wingo, then the director of RFF's Regional and Urban Studies Program, negotiated the agreement with the academy which made this volume possible. John Jackson served ably as executive secretary for the committee, and was responsible for general arrangements. Finally, of course, we must thank Wallace Oates who was the organizer and editor of this volume.

Marion Clawson
Vice President
Resources for the Future, Inc.

xiii

1 Introduction

WALLACE E. OATES

Late in 1972, the U.S. Treasury Department mailed checks totaling more than $2.5 billion to 37,000 state and local governments, including all of the states and virtually all county, municipal, and township governments in the nation. Revenue sharing, following a tortuous legislative history, had arrived.

As an integral part of the Republican program for a New Federalism, revenue sharing embodies an important shift in political power and responsibilities: a movement to decentralize fiscal decision making in the U.S. federal system. Federal forms of governments, by their very structure, are characterized by a continuing tension among their various levels. At the topmost level, the central government constitutes a concentration of power capable of effecting change on a national scale and reaching beyond purely regional concerns; much of the twentieth century has, in fact, been a process of growing centralization of the public sector to cope with, among other things, sometimes massive fiscal dislocations.

Inevitably, disillusionment with what will always be an imperfect world calls into question existing political trends. Since many centrally directed programs to remove poverty, to ease the ills of the cities, and to provide improved health care have not fully resolved the problems, criticism of these programs has mounted. The "distant" central bureaucracy is a convenient scapegoat. Moreover, decentralization of the public sector has, in principle, much to commend it: the proximity of state and local governments to the people and their problems can promote a better understanding and responsiveness to local needs; local government may encourage greater participation and concern by the citizenry; and the many distinct political units can provide a range of experimentation with different solutions to

1

existing social and economic problems. The states should be, as Justice Brandeis observed, the "laboratories of the federal system."

With the advent of revenue sharing, the pendulum seems to be swinging back to a more decentralized fiscal structure.[1] In view of the origins of the proposal for revenue sharing, this is somewhat ironic. In the mid-1960s, Walter Heller and Joseph Pechman had urged the adoption of unconditional federal assistance to state and local governments, but their primary motives were a good deal different from those that eventually proved decisive.[2] Heller, in particular, saw in revenue sharing a solution to two apparently unrelated problems. The first was the depressing effect that the automatic growth of federal tax revenues had on the economy. While the Revenue Act of 1964 (involving a large tax cut in excess of $10 billion to individuals and corporations) had temporarily injected needed purchasing power to stimulate a sagging economy, Heller saw a continuing problem of "fiscal drag" generated by the tendency of federal tax collections to expand more rapidly than the economy (or the need of then existing federal programs). If no offsetting actions were forthcoming, the federal budget could be expected, over time, to generate an increasingly contractive impact on the aggregate level of economic activity.

Second, the decade of the 1960s was a period of intense fiscal pressure on state and local governments. Rapidly expanding school enrollments, growing demands for urban services, and soaring unit costs had induced enormous increases in state and local expenditure. Pechman, quite appropriately, observed that "the largest 'growth industry' in the United States has been state and local government."[3] Moreover, state and local tax structures were far less income-responsive than was the tax system of the federal government; the built-in growth in state–local tax collections was simply inadequate to meet budgetary commitments of existing expenditure programs so that the vast majority of jurisdictions were forced to introduce new forms of taxation and increases in tax rates. In brief, Heller described the U.S. structure as a "fiscal mismatch": the federal government controlled the income-elastic revenue sources, while the state and local governments were saddled with the rapidly expanding expenditure functions.

---

[1] This, incidentally, appears to be true not only of the United States, but elsewhere as well. The fiscal accounts of most of the industrialized countries indicate a declining share for the central government over the last two decades. See W. Oates, *Fiscal Federalism* (New York: Harcourt Brace Jovanovich, 1972), chapter 6.

[2] Heller, *New Dimensions of Political Economy* (Cambridge: Harvard University Press, 1966); and Pechman, "Financing State and Local Government," The American Bankers Association, *Proceedings of a Symposium on Federal Taxation* (Washington, D.C.: 1965), pp. 77–84.

[3] Pechman, "Financing State and Local Government," p. 72.

Revenue sharing seemed to offer a happy, simultaneous resolution of these two problems. The federal government would, in effect, correct this mismatch by making available to state and local governments a large chunk of the *increment* to federal revenues in the form of unconditional grants-in-aid. This would serve both to bolster aggregate demand for the economy at large and to channel resources to a key (and badly hurting) sector of the economy.

While Heller and Pechman saw revenue sharing as providing a new source of support and vitality for state and local governments, they did not envision a major shift in the balance of power in the U.S. federal system. In particular, revenue sharing was not to come at the expense of conditional aid. In Heller's words, "As we parcel out future fiscal dividends, grants-in-aid will be near the head of the queue. Conditional grants for specific functions play an indispensable role in our federalism."[4] Matching provisions on conditional grants, for example, would continue to provide needed inducements to state and local officials to expand services and programs that the federal government thought would otherwise receive inadequate support. Revenue-sharing funds would supplement, not supplant, existing conditional aid.

To avoid annual legislative action and thereby insure continuity of funding, Heller and Pechman recommended that revenue-sharing payments come from a trust fund whose source would be a specified percentage of the federal personal income tax base. In their original proposal, the dollars were to be distributed on a per capita basis to the states, with no strings attached aside from some approved plan for the pass-through of funds to local governments. Heller emphasized that "the federal commitment to share income tax revenues with the states would be a contractual one in the sense of being payable—in whatever percentage Congress provided— through thick and thin, through surplus and deficit in the federal budget."[5]

The original vision of revenue sharing was that of a relatively simple, almost self-administering program that would both provide continuing growth in aggregate demand and extend needed assistance to hard-pressed state and local governments. Much would happen to this proposal, however, before it was finally enacted by the Congress.

When the Republicans turned to revenue sharing in the late 1960s, it was with a considerably different emphasis: it became a principal weapon in their attack on President Johnson's Great Society. The failure of the Johnson programs was, the Republicans contended, the result of trying to impose solutions from the nation's capital. The newly elected President

---

[4] Heller, *New Dimensions*, p. 141.
[5] Ibid., p. 148.

Nixon, in his revenue-sharing message, expressed concern over the growing power of the federal government and noted, "Yet, despite the enormous Federal commitment in new men, new ideas and new dollars from Washington, it was during this very period in our history that the problems of our cities deepened rapidly into crises."[6]

Revenue sharing was now to become an essential element in a process of decentralizing power in the U.S. federal system. By 1971 Mr. Nixon had elevated the proposed program to one of his "six great goals" for the nation. In his State of the Union Message on January 22, 1971, the President emphasized that:

> The time has come now in America to reverse the flow of power and resources from the States and communities back to Washington, and start power and resources flowing back from Washington to the States and communities and, more important, to the people all across America.
>
> . . . So let us put the money where the needs are. And let us put the power to spend it where the people are.[7]

With the rapid increase in federal expenditure necessitated by the Vietnam war, supplemented by domestic spending on programs for the Great Society, the fiscal dividend had long since disappeared, and little more was to be heard of the problem of fiscal drag. Moreover, there was considerable discontent in some quarters with the rapid growth and fragmentation of many conditional grant programs. In particular, the federal government during the 1960s had tried to attack poverty, discrimination, and urban decay, in part by circumventing traditional government units with such things as the Community Action Program. By channeling funds directly to often-dissident, minority groups, federal officials hoped to move resources into the hands of those who could and would deal directly with the ills besetting urban society.

Such tactics naturally generated resentment among some governors and city mayors, as they found themselves subject, in some instances, to a federally supported opposition. In revenue sharing the Republicans found an ideal program to capitalize on this discontent. Not only would revenue sharing serve to decentralize resources and power, but it would work through the traditional structures of state and local governments. Rather than being merely a supplement to existing conditional grant programs,

---

[6] Message from the President to the Congress, Aug. 13, 1969, *Weekly Compilation of Presidential Documents* (Washington, D.C.: Office of the Federal Registrar, Aug. 18, 1969), p. 1143.

[7] The State of the Union Message, Jan. 22, 1971, *Weekly Compilation of Presidential Documents* (Washington, D.C.: Office of the Federal Registrar, Jan. 25, 1971), pp. 92–94.

revenue sharing provided the occasion for a comprehensive review of the entire system of federal aid. For their part, many Republicans saw in all this the opportunity to dismantle many of the Great Society programs which had displeased them.

The proposal came to the Congress in two distinct parts. The first was for a general revenue-sharing fund consisting of monies to be distributed with virtually no strings attached to state and local governments. In addition, the administration proposed the consolidation of some 130 categorical aid programs into six bloc grant programs to deal with major domestic problems: urban community development, rural community development, manpower training, education, law enforcement, and transportation. These would constitute special revenue sharing under which funds would be designated for a broad area of programs, but without any matching requirements or detailed stipulations, so that here also recipient units would be able to substitute their priorities for those of Washington.

The passage through the Congress of the revenue-sharing proposal was predictably not an easy one. There were many competing interests seeking federal support, and federal legislators were reluctant to turn over large sums of federal money to state and local units with no federal controls over their use. In particular, many congressmen balked when it came to special revenue sharing. Although they admitted shortcomings in existing conditional grant programs, they saw these proposals as throwing out the good programs along with the bad. Not surprisingly, the special revenue-sharing proposals have made slow progress through the Congress; at this date, only two special revenue-sharing bills have actually been enacted—the Comprehensive Employment and Training Act of 1973 and the Housing and Community Development Act of August 1974.[8]

General revenue sharing itself underwent a considerable metamorphosis in the legislative branch. Wilbur Mills, then chairman of the House Ways and Means Committee, opposed, among other things, the dilution of congressional direction and review of the spending of federal funds; he was especially against complete discretion for state and local governments in the use of revenue-sharing monies and the circumvention of standard appropriation procedures. In the end, Mills prevailed on most counts. The House bill inserted into the revenue-sharing formula provisions which would make a state's entitlement depend not only on its population and tax effort (as proposed by the administration), but also on the relative size of its urbanized population, its level of per capita income, and its relative reliance on income taxation. In addition, Mills insisted that local

---

[8] In addition, some new law enforcement and education legislation has incorporated certain ideas from the original special revenue-sharing proposals.

governments spend their revenue-sharing monies on certain high-priority programs: public safety, transportation, health, environmental protection, and a few others. Finally, the House refused to establish a trust fund; it decided instead to vote the usual congressional appropriation with an initial life of five years.

To complicate matters further, the Senate found it could not support the House formula for the allocation of revenue-sharing funds and so proposed its own. In the end, both formulas were incorporated into the revenue-sharing bill, and each state was allowed to choose that formula which gave it the largest sum.

Perhaps most striking is that what originated as a relatively straightforward proposal for unconditional support for state and local governments has emerged as a highly complex fiscal program with a whole myriad of incentives, many of which the framers of the legislation most certainly did not recognize. In fact, the newly emerging revenue-sharing programs contain a potentially quite powerful set of forces operating on the U.S. political and fiscal system.

It is, therefore, of great importance that we seek to understand what we have done, and are doing, in the course of modifying the structure of intergovernmental fiscal relations in the United States. This is the objective of the studies in this collection. The first two papers—those of Jeffrey L. Pressman and Robert D. Reischauer—address directly the political and fiscal implications of revenue sharing.

What both these papers make clear is the complicated, often largely hidden, sets of incentives embedded in the revenue-sharing bills. In fact, the most explicit and seemingly most important "strings" in the provisions may turn out not to be binding constraints at all. Local governments, for example, are required to expend their revenue-sharing allocations on certain high-priority functions (and not to use them for local tax relief). However, as Pressman points out, this stipulation is rather unlikely to restrict local governments in their fiscal decisions. Particularly in times of expanding budgets, funds are highly fungible; it is easy for local officials to use revenue-sharing funds to meet mandated increases in, say, salaries of policemen and firemen, in which case the revenue-sharing grant could simply serve to hold local taxes to a lower level than would be possible in the absence of these funds.

It requires careful analysis, and in some instances a bit of guesswork, to determine what the real effects of revenue sharing are likely to be. Pressman's chapter explores the political implications, over the short and longer term, of both general and special revenue sharing. Certain facts are clear. Revenue sharing does promise to alter the fiscal resources at the disposal of different governmental units, and this implies shifts in the structure of power. More revenues will flow into the treasuries of state and local governments. This has important implications for the balance of power between the Congress and federal bureaucracy, on the one hand, and decentralized public units, on the other. Pressman points out, "It is not surprising that members of Congress have been among the most vocal critics of proposed shifts from categorical programs to discretionary funding, because these changes pose threats to certain long-standing congressional prerogatives." Similarly, those federal officials seeking a "social activist role" fear that their power might be reduced to a check-signing function.

Moreover, as noted earlier, revenue sharing promises to strengthen the hold of traditional governments at the expense of many relatively new and innovative "paragovernments." Since only general purpose governments are eligible for revenue-sharing funds, we can expect to see less effort and fewer resources directed into the formation and operation of the many special authorities that have come into being in the United States. Pressman is careful to point out, however, that the special expertise developed in many of the independent agencies will continue to be valuable to city government; many city development agencies, for example, possess a technical capacity that the mayor and his staff do not have at their disposal. With less funding, there is thus some uncertainty as to the future of many of these governmental units.

How can the recipients be expected to employ their revenue-sharing funds? It is too early to say anything definitive on this, but Pressman's own research and that of others suggest that the early experience has not been one of initiating new and innovative programs, but rather that of financing the usual services. While this may seem hardly surprising, caution is in order, for, as Pressman notes, "the ease of moving money between accounts" makes it extremely difficult to interpret the budgetary data.

Reischauer's chapter contains a thorough (and badly needed) analysis of the House and Senate formulas for general revenue sharing. This will provide a number of surprises (and shocks) to the authors of the bill. The complex of largely arbitrary and unintended effects concealed in the formulas and accompanying provisions is, at the least, quite disturbing. One striking anomaly emerged when Reischauer simply determined how the per capita entitlements of the fifty states and the District of Columbia differed under the House and Senate formulas: the simple correlation is negative (that is, those states which *on average* would receive relatively large per capita payments under the House formula would, in contrast, have comparatively small entitlements under the Senate formula).

With great care and detail, Reischauer moves through the effects of the various terms in the formulas themselves and how they interact with the other stipulations of the bill. How much additional revenue-sharing aid can a state obtain by increasing its income tax collections by $1.00, or how much can a local government get by raising its tax effort? Reischauer provides state-by-state estimates of the magnitudes of all these incentives. While, in many instances, the implied subsidy is quite small, in other circumstances (if, for example, the government is operating at one of the specified floors or ceilings) the incentive can be a very powerful one. As an illustration, the state of Ohio in 1972, by increasing its income tax receipts and reducing state sales taxes by a like amount could have increased the state's revenue sharing allocation by 35 percent; in contrast, Minnesota would have obtained no increase at all, while the state of New Jersey would receive no additional assistance unless it greatly expanded its income tax collections.

Some of the implications for governmental structure are likewise quite provocative. The guarantee that all general purpose governments receive at least 20 percent of the average statewide grant per capita may save many obsolete, "do-little" governments whose responsibilities have been transferred to overlying government units. As Reischauer states, some of these "atrophying units . . . have developed into little more than special purpose districts that maintain a few miles of local roadways or carry out functions explicitly defined and financed by the state government." Revenue sharing makes it in the interest of local inhabitants to preserve the existence of such units, even though administrative efficiency may call for their elimination. Among the local governments that benefit from this floor "are 423 townships and villages in New York, 1,038 in Ohio, 705 in Wisconsin, and 870 in Minnesota."

Reischauer systematically enumerates and evaluates the lengthy list of incentives built into general revenue sharing. In each case, he not only describes the incentives inherent in a particular provision, but wherever possible presents quantitative evidence of the magnitude of the incentive and the likely response to it. In particular, he looks at the experience over the first year of revenue sharing to see if he can discern any patterns of responses. For the most part, he is unable to do so, but this is hardly surprising. First, there has not been sufficient time to effect significant fiscal or institutional change. Second, the uncertainty over the future of revenue-sharing programs themselves may be enough to make public officials delay the introduction of new measures. And third, as Reischauer's essay itself demonstrates, many of the incentives are so subtle that it may take some time for state and local governments to realize just how to respond. The Reischauer paper should itself help to clarify this.

In addition to the implications of newly enacted revenue-sharing legislation, there is the larger issue of the whole structure of intergovernmental fiscal relations in the U.S. federal system. Conditional grants have historically been a fundamental element in this system, and there remains, in principle, a strong case for a continuing role for various types of conditional aid. The really tough question is what overall composition of grant programs would best promote our social objectives. To try to answer this question, we require a variety of information. First, we obviously need to know what our objectives are. This is not easy, for our society is composed of many individuals and groups with differing interests, who, through the political process, must somehow provide us with a social ordering of priorities.

Second, we must know how the system works. For example, if we increase the federal matching rate on a public assistance program, by how much are local authorities likely to expand payments to the recipients of the program? For a successful application of the "social engineering" process, we must determine how the social system behaves and, in particular, how it responds to the various policy instruments we have at our disposal.

The chapters by Robert P. Inman and Martin McGuire explore this set of issues. Inman's procedure is to set forth a general equilibrium model of the economic system, to enumerate a number of different social welfare functions, and then to experiment quantitatively with differing programs of grants-in-aid to see which programs score highest under each of the social welfare functions. His subject for the quantitative analysis is aid for public schools.

Inman's analysis is admittedly of an exploratory and illustrative character but, tentative though it may be, it is illuminating on several critical matters. First, Inman's insistence on a general equilibrium framework is to be emphasized. The outcome of a new public policy depends not only on its initial impact on the system, but also on the sequence of responses throughout the system as conditions change. For example, it is not enough, when analyzing a program of school aid, simply to add the aid to the existing tax base. The introduction of the aid program will itself change the size of the original tax base. In fact, Inman finds such capitalization of the fiscal flows into property values to be an important determinant of the ultimate allocative and distributive effects of alternative grant programs; the comparative effects, for instance, of foundation and matching aid programs depend significantly on the differential responses of local tax bases to the two programs.

Second, we have suggested previously that the choice of policy instruments may depend on the particular values of the parameters of the system. Inman's model indicates that the effects of different school-aid policies can be crucially dependent on such characteristics of the system as the price elasticity of local demand for education. Inman's finding, for example, of the relative ineffectiveness of district matching aid is largely attributable to the low price elasticity of demand in his sample metropolitan economy.

And third, Inman's findings remind us of the frequent conflicts between the objectives of economic efficiency and equity. A program which equalizes expenditures per pupil may get high marks on equity grounds, but it is likely to involve allocative losses from distorted or suppressed individual choice.

Martin McGuire is also concerned with the response to different grant-in-aid programs. His approach is an intriguing one. Normally, economists define an objective function for the decision maker, enumerate the constraints, and then proceed to trace out the behavioral patterns implied by utility maximization subject to the specified constraints. There are at least two major difficulties in employing this procedure to study intergovernmental grants. The first is that such grants are not transfers to individuals; they are transfers to governmental units or groups of people. Furthermore, the response to the grant presumably depends on the political institutions (or rules for collective choice) of the recipients. An individualistic objective function together with an assumption of rational utility maximization may not provide a very apt description of this political response. This becomes a particularly difficult issue when a new program like revenue sharing is introduced. As we noted earlier, revenue sharing may bring some substantial shifts in the structure of power; it may change the relative influence of those whose tastes count in the determination of public outputs (for example, by shifting spending decisions to governors and mayors away from paragovernmental units). As such, programs like revenue sharing may alter the system in a way that most closely resembles a change in the "tastes" of the decision maker. To consider it as simply altering the budget constraint may be a misleading simplification.

On this first matter, we really have not made a great deal of headway[9]; McGuire simply assumes that, whatever the choice mechanism, it can be analyzed as a systematic response to price and income effects into which all grants can implicitly be partitioned.

The second difficulty is that the great maze of existing grant programs (some matching, some not; some open-ended, some closed) makes it extremely difficult to specify the actual budget constraint confronting state

---

[9] For one effort in this direction, see D. Bradford and W. Oates, "The Analysis of Revenue Sharing in a New Approach to Collective Fiscal Decisions," *Quarterly Journal of Economics* (August 1971), pp. 416–439; and their "Towards a Predictive Theory of Intergovernmental Grants," *American Economic Review* (May 1971), pp. 440–448.

and local governments. McGuire's technique is to assume for analytic purposes that the constraint itself is unknown to the researcher and then, by introducing some additional restrictions on the form of the utility function, to let the observed data tell us what the constraint is as well as the recipient's response to it. This is a provocative approach to an extremely troublesome obstacle in determining the effects of intergovernmental grants, and McGuire is able to make effective use of it. In particular, he sets forth a menu of models for the measurement of the budgetary response to intergovernmental grants. Unlike earlier studies, these models permit us to distinguish empirically between the income and price effects implicit in the grant. It is this kind of information that is essential to the design of an effective system of intergovernmental aid; the prospect of some actual econometric estimates of these models by McGuire (and perhaps others) is thus an exciting one. At present, it is intriguing in his chapter to see an innovative approach to the determination of the fiscal effects of intergovernmental grants.

The last paper in the volume comes back to one of the original tenets of revenue sharing: the alleged fiscal mismatch between the federal and state–local governments. Proponents of revenue sharing (and of state–local tax reform as well) have contended that a major source of the fiscal distress of decentralized governments is the relatively income-inelastic structure of their tax systems; revenues simply do not grow sufficiently rapidly at existing tax rates to keep pace with needed increments to the budget.

This whole argument depends on an implicit (and, as yet, largely unexamined) proposition: the expansion in the public budget depends, other things being equal, on the income-elasticity of the tax structure. While this hypothesis has a certain pragmatic appeal, I argue that its validity is far from obvious. Upon closer examination, this assumption seems to imply some very curious behavior on the part of taxpayer–voters. Specifically, it suggests that they will support expansions in the budget if they can be financed with no increase in tax rates (that is, by increments to revenues coming from growth in the tax base), but they will oppose this same spending if it requires hikes in tax rates (or the introduction of new taxes). In short, the proposition seems to imply that what people care about are their tax *rates*, not their tax *bills*—hardly the sort of behavior we normally attribute to rational consumers.

The point is that the proposition itself is a testable hypothesis and one that may well be untrue. I set out to test this hypothesis by a cross-sectional regression analysis of the behavior of state budgets and of city budgets over the decade 1960–70. After controlling for the effects of other variables, I found that my measures of the tax-elasticity of the tax system do seem to have had a statistically significant, although not quantitatively very large, effect on the growth in public spending. The impact would appear hardly large enough by itself to justify major fiscal reforms.

This does not, of course, imply that revenue sharing is a total chimera. It does effect some redistribution of income among jurisdictions, and, even if revenue-sharing funds are simply substituted for locally generated revenues, we might find that an overall fiscal structure with heavier reliance on federal revenue sources would generate less inefficiency in resource use and a more desirable pattern of tax incidence. What the analysis does call into question is that revenue sharing will generate substantially higher levels of state and local expenditures. Why should a citizen care whether the dollar he pays in taxes to support local public services goes directly into the local treasury or, alternatively, passes to the central treasury to be returned to his local jurisdiction? In either case (assuming a costless transfer process), he has paid a dollar to finance local services.

The chapters in this volume thus provide both a critical evaluation of the recently enacted revenue-sharing bills and a broader analysis of possible designs for a more effective system of intergovernmental grants-in-aid. At this juncture in our history, there remain many crucial and unanswered questions concerning the future of our federal fiscal structure. What seems clear is that we can expect a continuing and profound interaction among different levels of government. Lord Bryce once described a federal system as resembling "a great factory wherein two sets of machinery are at work, their revolving wheels apparently intermixed, their bands crossing one another, yet each doing its own work without touching or hampering the other."[10] This view is clearly outdated. Modern fiscal federalism is characterized by a continuous overlap of programs with activities at different governmental levels cutting sharply across functional lines; Lord Bryce's revolving wheels are now intermeshed, and intergovernmental grants promise to play a central role in this structure. Studies of the kind presented in this collection should help us to understand more adequately how these grants actually influence the operation of the public sector and how we can best employ them to promote an efficient use of resources and a just distribution of income in our society.

[10] Taken from Geoffrey Sawer, *Modern Federalism* (London: C. A. Watts, 1969), p. 66.

2 Political Implications of the New Federalism

JEFFREY L. PRESSMAN*

Although it has been much discussed, praised, and attacked, the *New Federalism* has frequently been defined so broadly that it is difficult to know what it means. The term has been applied to a wide range of governmental actions: revenue sharing, reorganization of the cabinet, cutting the funding of certain domestic programs, decentralizing executive agencies, strengthening the management component of the Office of Management and Budget (OMB), and reform of the welfare system, to name but a few.

In this chapter, I will use the term *New Federalism* to refer to its central component: the replacement of federal categorical programs by relatively discretionary grants to state and local governments. I will focus specifically on general revenue sharing (the provision of virtually unrestricted federal tax revenues to states and localities), special revenue sharing (the joining of categorical programs into broad, functional bloc grants), and certain federally sponsored transitional measures which have been designed to prepare local governments for the shift in funding patterns.

A Focus on the Political Dimension

The decentralization of responsibility for operating federal programs is not merely a technical administrative exercise, for by changing the pattern of distribution of federal funds and by changing the governmental machinery by which those funds are distributed, the New Federalism will affect the

*Associate Professor of Political Science, Massachusetts Institute of Technology.

distribution of political influence—the capacity of individuals and groups to achieve their goals and to influence others to act according to their wishes. Altering the structure of federal grant programs will increase the political resources of certain groups and institutions while decreasing the resources of others, thus altering the power relationships within the political system. (As examples of "political resources," I would include money, authority, information, and administrative capacity.)

Although changes in federal programs have implications for the distribution of political resources both between and within levels of government, this relationship is not one-way. The outcomes of these program changes will be strongly influenced by existing institutional characteristics, resource distributions, and patterns of behavior. Changes in federal delivery systems, no matter how carefully designed, cannot abolish political relationships which have developed over time. In discussing the political dimension of the New Federalism, this chapter will focus both on the implications of the federal changes for existing political relationships and on the ways in which those relationships may influence the outcomes of the new federal policies.

It is important to note that both proponents and critics of the New Federalism have placed great emphasis on the political implications of changes in the aid system. Supporters of increased discretionary funds for state and local governments have argued that such funds would have a revitalizing effect on those governments. Walter W. Heller, an early proponent of the idea of revenue sharing, declared, "Transcending all other considerations, as we seek new forms of Federal fiscal relief for the states, is the need not simply to increase their resources but to restore their vitality; not simply to make them better 'service stations' of federalism but to release their creative and innovative energies."[1]

In his State of the Union Message in 1971, President Nixon used ringing rhetoric to outline a political rationale for New Federalism programs:

> If we put more power in more places, we can make the government more creative in more places. . . . The further government is from the people, the stronger government becomes and the weaker people become. And a nation with a strong government and a weak people is an empty shell. . . . Local government is the government closest to the *people*, it is most responsive to the individual *person*. [The emphasis is President Nixon's.]

The President ended his speech by telling the Congress that revenue-sharing programs could open the way to a "New American Revolution—a peaceful

---

[1] Walter W. Heller, *New Dimensions of Political Economy* (Cambridge, Mass.: Harvard University Press, 1966), p. 168.

revolution in which power was turned back to the people—in which government at all levels was refreshed and renewed, and made truly responsive."[2]

Opponents of New Federalism programs have also pointed to the political implications of those programs. They have argued that the removal of federal controls will threaten efforts to redistribute resources to people who now lack them—poor people and racial minorities. Wilbur J. Cohen, a former secretary of health, education, and welfare, has argued, "We have to have federal programs with strings attached because it is the only way that the disadvantaged, the poor whites and poor blacks will get their fair share. If there are not federally regulated programs to disburse money and instead it is handled by local city governments, then they won't get their fair share."[3]

The controversy surrounding the New Federalism and its political impact may be seen as part of a long-standing debate in this country between advocates of centralization and advocates of decentralization. This debate has been closely watched (and occasionally entered into) by political scientists interested in the influence of the size of a political unit on the kinds of political behavior occurring within the unit.

After identifying some of the issues involved in past discussions about the appropriate size of a political unit, I will move to an examination of particular New Federalism programs and the political questions they have raised. Then I will explore some of the underlying, long-range implications of a shift from national to local authority, and will speculate on the ways in which the workings of our federal system may be changed.

The Issue of Size

One of the oldest and most often repeated questions of political theory is that of the appropriate unit for a democratic political system. Those who have grappled with this question have found it to be a difficult one. Robert A. Dahl has written that, in a world of high population densities and great interdependence, the search for an appropriate democratic unit must confront this basic dilemma: the larger a unit is, the more its government can regulate those aspects of its environment that its citizens want to regulate. But the larger a unit is, too, and the more complex its tasks, the less direct

---

[2] The State of the Union Message, Jan. 22, 1971, *Weekly Compilation of Presidential Documents* (Washington, D.C.: Office of the Federal Registrar, Jan. 25, 1971), pp. 93, 94, and 97.

[3] Cohen, quoted in "The New Federalism: Theory, Practice, Problems," *National Journal* special report (March 1973), p. 14.

participation will be available to its citizens. Conversely, the smaller a unit is, the greater the chance for citizens to participate in governance, but the range of control of the environment is less. Thus, for most citizens, Dahl argues that participation in very large units is minimal. But participation in small units, though much more extensive, runs the risk of being trivial.[4] Therefore, the search for an appropriately sized democratic unit involves difficult tradeoffs between the intrinsic quality of participation and the effects of that participation on the outside world.

Of course the discussion of unit size has not taken place in a political vacuum. Advocates of both centralization and decentralization have recognized that the size of a political unit can influence the kind of behavior that goes on within that unit and the kinds of groups who will be advantaged or disadvantaged. The size and shape of a constituency can have an important effect on the decisions that will come out of that constituency.

The belief in the virtues of decentralization and small political units, which is evident in the statements of New Federalism supporters, has been a continuing theme of American political thought. Proponents of small units have asserted that such units are more natural, that they are closer to the people, and that they facilitate rational decision making by allowing citizens to discuss their affairs in a face-to-face manner. The government of a small unit can, it is said, easily ascertain citizens' policy preferences, and the citizens are close enough to their government to be able to exert control over it.

Although the traditional models of small communities in America have been rural towns (with inevitable references to the New England town meeting), certain urban groups have in recent years espoused the arguments for decentralization as well. Black activists, pressing for "community control" of municipal services, have set forth the often cited virtues of participation, face-to-face discussion, and increased governmental responsiveness. It is important to recognize, however, that different proponents of "decentralization" are often advocating very different policies. Conservatives have traditionally called for the granting of more authority to state and local officials, while black urban groups have advocated the lodging of power in neighborhood groups (with continuing resource support by the national government).

Critics of decentralization have argued that small units are more susceptible than larger ones to domination by a single group. James Madison made that point in the following passage from *The Federalist:*

[4] Dahl, "The City in the Future of Democracy," *American Political Science Review*, vol. 61 (December 1967), pp. 953–970.

The smaller the society, the fewer probably will be the distinct parties and interests composing it; the more frequently will a majority be found of the same party, and the smaller the compass within which they are placed, the more easily will they concert and execute their plans of oppression. Extend the sphere, and you take in a greater variety of parties and interests; you make it less probable that a majority of the whole will have a common motive to invade the rights of other citizens; or if such a common motive exists, it will be more difficult for all who feel it to discover their own strength, and to act in unison with each other.[5]

Thus, for Madison, the diversity that characterizes larger political units meant that no single interest could dominate political life in those units.

This Madisonian argument for large-sized units has been a foundation for later writers who have preferred federal to local power. Grant McConnell, a supporter of strong central government, has asserted that "decentralization will generally tend to accentuate any inequality in the distribution of power that would otherwise exist *within* each decentralized unit."[6] In larger communities, McConnell has asserted, impersonality and diversity actually facilitate access to decision making. But in smaller units, informal social pressures suppress any challenge to the dominant group; as a result, the dominant group, freed from the competition it might have to face in a larger unit, tends to be the most powerful economic interest in a particular locality. McConnell asserts that reliance on state and local governments for decision making means that more broadly based public interests will lose out to narrow private interests. Other losers will be minority groups which, if distributed evenly among numerous small constituencies, will lack the numbers necessary to achieve any influence at all within those constituencies.

Contemporary arguments over the New Federalism repeat recurrent themes in American political history upheld by advocates of centralization and decentralization. Proponents of decentralized decision making stress the virtues of grass-roots politics: local control, participation, firsthand knowledge of local conditions, and freedom from dictation by the central government. On the other side, critics of New Federalism's decentralization talk of the necessity of maintaining national standards, and warn that "public interest" goals and minority group rights may be endangered if decision-making authority is turned over to local governments.

---

[5] *The Federalist*, no. 10 (New York: Modern Library), pp. 60–61.

[6] McConnell, *Private Power and American Democracy* (New York: Knopf, 1966), p. 107.

The debate over the New Federalism has not, of course, been an isolated discussion of political theory, for many of the participants in that debate have been groups whose strengths are differentially distributed among the levels of government. Labor unions, racial minority groups, and big-city mayors, for example, have achieved their greatest degree of influence at the national level; they are usually found in opposition to proposals to turn programs over to the states. But rural and suburban interests, who feel that they may gain under New Federalism, have supported decentralization. Thus, as is often the case, preferences about governmental structure reflect preferences about substantive outcomes and particular group interests. The question of preferred level of government is inextricably linked to the configurations of influence at each level.

Having placed arguments about the New Federalism in the context of the continuing debate about political unit size, let us move now to a consideration of the political questions raised by particular New Federalism programs. The focus will be on the effects of these programs on governmental structures, processes, and the distribution of political resources. It will also be shown, however, that existing political relationships can themselves have an effect on the ultimate outcomes of the new programs.

New Federalism Programs and Their Impact

General Revenue Sharing

On October 12, 1972, Congress gave final approval to a bill providing virtually string-free payments to general purpose governments at the state, county, and city–town levels. President Nixon signed this bill, the State and Local Fiscal Assistance Act, on October 20, and in December the first checks were mailed to the more than 38,000 general purpose governmental units in the United States. During the period from January 1972–December 1976, this program of general revenue sharing will distribute $30.5 billion to state and local governments. (Robert Reischauer provides an analysis of the distribution formulas of the program in Chapter 3.)

The Strings That Remain

Although I have characterized revenue-sharing funds as being virtually string-free, it must be pointed out that the program does include some formal restrictions on the use of the money. Specified accounting procedures must be followed; programs funded by revenue sharing must be nondiscriminatory; construction projects which receive at least 25 percent of their funds from revenue sharing must follow Davis–Bacon wage guide-lines; the uses of revenue-sharing money must be published; and recipients cannot use the money as matching for other federal grants.[7] Furthermore, although state governments are permitted to spend revenue-sharing funds on any program they select, local governments are directed to use the money for certain "priority expenditures."[8]

In many instances, these formal strings may not prove to be very restric-tive. As Reischauer has convincingly argued, those state and local officials with expertise in budgeting and grantsmanship should find little resistance if they want to circumvent revenue sharing's administrative constraints.[9] For large cities and active county governments, the revenue-sharing entitle-ment represents only a small fraction of current program spending. Such jurisdictions will be able to slip their revenue-sharing grant into an existing program and then use the freed resources as they wish—for tax relief or other nonpriority expenditures. For strategic reporting purposes, local officials can list revenue sharing as having been used for areas in which the required accounting, nondiscrimination, and Davis–Bacon provisions were already in effect. In large jurisdictions, therefore, the strings on revenue sharing may turn out to be close to no strings at all. But as Reischauer points out, program restrictions may have a significant effect on the be-havior of officials in smaller units of government. The size of the revenue-sharing grant is large relative to the total operations of these units, and the narrow range of their activities will make it harder to shift funds between various accounts.

The Impact

Because strategic moves will make it difficult to trace revenue-sharing expenditures, and because it is hard to guess what expenditure patterns would have been in the absence of revenue sharing, it will not be easy to assess the substantive impact of general revenue sharing. This will doubtless prove frustrating to both proponents and critics of the New Federalism, for members of each group would like to point to evidence to support their views. Defenders of revenue sharing have maintained that it will lead to innovative expenditure decisions at the local level while opponents of the

---

[7] For an enumeration of these restrictions, see the *State and Local Fiscal Assistance Act of 1972*, Subtitle B: "Administrative Provisions."

[8] Ibid., sec. 103: "Use of Funds by Local Governments for Priority Expenditures."

[9] See Robert D. Reischauer, "On Evaluating General Revenue Sharing" (Brookings Institution draft, October 1973), pp. 50ff.; as well as Chapter 3 of this volume.

program have predicted that funds will be used for frivolous expenditures, or will be put into those interests already dominant in local communities. Although there have been scattered press reports about the use of revenue-sharing funds for building extra golf courses and monuments on the one hand, and for innovative social programs on the other, the evidence so far demonstrates neither widespread waste nor widespread innovation. A survey of cities over 50,000 in population, carried out in the spring of 1973 by David Caputo and Richard Cole, found that most cities reported spending their revenue-sharing funds in a pattern similar to that of their "normal" spending decisions.[10] The survey did not find much inclination on the part of the cities to spend revenue sharing for social programs. Only 1.6 percent of the new funds were spent for social services for the poor and aged, and only 1.1 percent for health services. (The authors note that a slightly smaller proportion of revenue-sharing funds has been allocated to the combined categories of health and welfare needs than that which would be expected in the normal budgetary process.) Although some administration officials have suggested that local governments might use their revenue-sharing money to pick up federal social programs, like the Office of Economic Opportunity (OEO) and Model Cities, which are being phased out, Caputo and Cole found that only a minority of the cities studied (19.8 percent) planned to use revenue-sharing funds for this purpose. It must be kept in mind, however, that this answer in the survey has tactical importance. If a city should report spending revenue-sharing money to make up for cuts in federal programs, it substantiates administration claims that the federal efforts are no longer needed.

In general, the study found that most cities applied their initial revenue-sharing funds to existing programs, rather than using the money to support new activities. This assessment was also reached by a later survey of 32,665 governmental units, carried out by the federal Office of Revenue Sharing (ORS).[11] (It should be noted that the federal report was also written by Caputo and Cole.) Such findings should caution us against expecting revenue sharing to result in widespread local innovation, at least not in the immediate future. It must be stressed again, however, that too much weight should not be given to formal statements by local officials on the uses of revenue sharing; the ease of moving money between accounts is considerable.

[10] Caputo and Cole, "The Initial Impact of Revenue Sharing on the Spending Patterns of American Cities." Paper presented at the annual meeting of the Southern Political Science Association, Atlanta, Georgia, 1973, p. 10.

[11] "U.S. Office Assesses Use of Revenue-Sharing Fund," *The New York Times*, March 1, 1974.

In any event, the revenue-sharing amounts are probably too small (well under half the recent annual growth in expenditures of recipient governments[12]) to cause major programmatic shifts at the local level. As a local government staff member remarked in the spring of 1974, "We used general revenue sharing to pick up current expenses. The amount of revenue sharing was roughly equal to the cost of inflation."[13] The switch from categorical programs to special revenue sharing, combined with alterations in funding levels, has the potential for causing significant institutional and programmatic change, but general revenue sharing is unlikely to accomplish this by itself.

The Process of Allocation

Going beyond the examination of the results of revenue-sharing allocations, it is also of interest to examine the processes by which those funds are allocated. In many localities, the process has been an internal one, dominated by the mayor, city manager, or city bureaucracy, and not including much citizen participation. This is not surprising, given the closed nature of urban budgetary processes.[14] But there have been a number of well-publicized cases of citizen involvement in the revenue-sharing allocation process: in Lakeland, Florida, and Tacoma, Washington, citywide polls were taken to determine how citizens thought their revenue-sharing money should be spent. In San Francisco, a series of public hearings resulted in a decision to fund neighborhood cultural centers. And in San Diego, a coalition of numerous neighborhood and other groups held a citywide convention to determine revenue-sharing priorities. Sometimes, the pressure for innovation comes from elected officials themselves, as in Seattle, where a city council majority drew up and passed a plan for spending revenue-sharing money on a range of new social programs.

Almost half of the cities in the Caputo–Cole survey reported having had public hearings prior to revenue-sharing decisions—and it is interesting to note that these hearings appear to have made little difference in reported expenditure decisions.[15] But one should view claims of open hearings and

[12] See Reischauer, "On Evaluating General Revenue Sharing," p. 51.

[13] As part of my research for this essay, I carried out open-ended interviews on the implications of the New Federalism with federal officials, congressional staff members, and local officials in Oakland, California, and Lowell and Boston, Massachusetts. In addition, I talked with representatives of national interest groups who have been active in the legislative struggles over general and special revenue sharing. The nonattributed quotes in this chapter are taken from these interviews.

[14] See Arnold J. Meltsner, *The Politics of City Revenue* (Berkeley: University of California Press, 1971).

[15] Caputo and Cole, "The Initial Impact of Revenue Sharing," pp. 12–15.

other forms of citizen participation with some caution. A number of questions might be posed: Were the hearings or decision sessions publicized? Were they held at a time when working people could attend? Were they held at a place accessible to all neighborhoods of a city? How many people actually participated? And what effect did the participation have on budgetary decisions?

In the coming years, it will be interesting to examine the ways in which state and local governments reach their revenue-sharing decisions, and to note the variations among these units. It may be that revenue-sharing funds, although limited in themselves, will become a focus around which citizens' groups will organize. Revenue sharing could be an entering wedge both for those who want to study the local political and budgetary processes and for those who want to influence those processes. Thus we should be careful not to disregard entirely the potential of general revenue sharing for causing political changes.

Governmental Structures

Another way in which revenue sharing could have a political impact is through the incentives it creates for changing the ways in which governments are structured.[16] Because only general purpose governments are eligible to receive money under the program, there is a bias against special districts (for education, water, sanitation, and a host of other local government functions). We might therefore expect to see fewer special districts created in the future, as well as pressures for the reabsorption of special districts by general purpose local governments.

Because of a statutory guarantee that all general purpose, local governments must receive at least 20 percent of the statewide per capita local grant, there is also an incentive to retain general purpose governments—even those that are obsolete and inactive. Examples of such inactive governments would include many of the 13,000 townships in the Midwest, which are frequently little more than road districts. Also in this category would be county units (in such states as Massachusetts, South Dakota, and South Carolina) that have few functions beyond operating a court system. The program's incentive to retain these kinds of governmental units poses problems for movements that have urged consolidation of local governments in the interest of efficiency.

Thus, although general revenue sharing will not cause a revolution in the pattern of local expenditures, the program does have certain implications for governmental organization and structure at the local level.

---

[16] For an extended discussion of these incentives, see Chapter 3 of this volume.

Special Revenue Sharing

The various special revenue-sharing programs may be expected to have a greater impact than general revenue sharing on the distribution of resources in the political system. Special revenue sharing involves the replacement of a range of major, federal, categorical programs—each having its congressional sponsors, administrative supporters, and local recipients—by broad, discretionary bloc grants to state and local governments. These program alterations appear to have the potential for changing a host of existing political relationships.

In 1971 President Nixon proposed special revenue-sharing programs in six areas: urban development, rural development, education, manpower, transportation, and law enforcement. These new packages were designed to consolidate categorical programs into general accounts that could be used in a flexible way by local governments; the oversight from Washington was to be sharply reduced.

The administration's hopes for quick legislative success on these proposals went largely unfulfilled, as congressmen—buttressed by interest groups who benefit from categorical programs—proved resistant to special revenue sharing. By July 1974, only one such bill (manpower) had been passed—and that bill represented a compromise with legislators who insisted upon writing certain federal guidelines into the final legislation. This did not mean that the New Federalism was dead or even dormant, however, as federal officials began to make some changes through administrative devices. Even before the new manpower bill passed Congress in December 1973, the Labor Department had taken administrative steps to distribute federal manpower funds according to a formula, rather than through project applications. (The reliance on formulas, rather than individual project applications, for distribution is a fundamental element in New Federalism programs.) Moreover, as will be discussed more fully later, the Department of Housing and Urban Development (HUD) has initiated a number of transitional programs designed to increase the control of elected local officials over its programs. Thus, although the legislative record of special revenue sharing is meager, the idea is still very much alive.

Manpower: A New Program and Some Changes in Resource Distribution

The first special revenue-sharing bill to pass, the Comprehensive Employment and Training Act of 1973, repeals and replaces the Manpower Development and Training Act (MDTA) of 1962 and the manpower training sections of the Economic Opportunity Act of 1964. Although a number of federal guidelines have been removed from the consolidated program, the new legislation does require that state and local governments submit their manpower plans to the secretary of labor for approval. The program distributes funds on a formula basis, rather than the project-application approach of categorical legislation; the formula is based on a locality's previous level of federal manpower aid, the number of workers unemployed, and a poverty index.

An examination of this program, in comparison with its predecessors, reveals changes both in the nature of local recipients and in the distribution of funding among localities. Under the new legislation, federal money will be channeled directly to state and local governments. All states, along with cities and counties of over 100,000 population, will be enabled to act as prime sponsors for manpower training programs. This represents a shift of authority to local, general purpose governments; in past categorical programs, the Department of Labor provided grants to a wide range of organizations (Community Action agencies, unions, corporations, and others, as well as local governments) to carry out manpower programs.

Changes in the federal delivery system are of great importance to local government recipients, who stand to have their supply of resources increased or reduced as a result of these alterations. Representatives of large cities have been particularly concerned about the impact of the new program. Estimates made for internal congressional use by the Department of Labor indicate that the formula distribution of manpower training funds will result in a substantial loss of such funds for cities over 100,000.[17] This is because many of the discontinued categorical programs—like the Concentrated Employment Program and Neighborhood Youth Corps—benefited mainly big cities. And some of the programs, like MDTA classroom training, which officially went to state governments, provided the bulk of their resources to the big cities where the training facilities were.

If the new program is giving some concern to big cities, it is being widely hailed by representatives of counties. Under the formula approach, suburban counties with populations of 100,000 or more will see their share of federal manpower funding increase significantly. Ralph L. Tabor, director of federal affairs for the National Association of Counties, has said, "When we consider where the counties were in manpower two years ago, this legislation represents a tremendous advance for us."[18] Critics of the new approach question the ability of county governments, which have limited

---

[17] See Robert Reischauer, "The New Federalism and the Old Cities: The Local Expenditure Implications of Shifting from Categorical to Block Grants" (Brookings Institution draft, December 1973), pp. 17ff., for an assessment of the impact of manpower special revenue sharing. Also see, "Manpower Report/Revenue sharing shift set for worker training programs," *National Journal Reports*, vol. 6 (Jan. 12, 1974), pp. 51–58.

[18] Ibid., p. 58.

experience in the manpower area, to shape effective programs. Representatives of counties and cities, recognizing that changes in federal programs have implications for who gets what at the local level, have struggled with each other to influence the shaping of the Manpower Act and other special revenue-sharing legislation.

Community Development Revenue Sharing: Competing Governmental Units and Philosophies

The form of special revenue sharing which has been of most interest to city governments has been the one earmarked for "community development." As it appeared in the administration's proposed "Better Communities Act," the new urban bloc grant would consolidate seven existing categorical aid programs. The largest of these programs is urban renewal, which has aided a thousand localities and distributed over $1 billion annually. Other large programs proposed for inclusion in the new bloc grant are Model Cities and water and sewer grants, with previous funding levels of over $500 million and $100 million, respectively. The remaining four programs are: open space and historic preservation, neighborhood facilities, rehabilitation loans, and public facility loans. Funds would, according to the administration's approach, be paid out automatically under a formula; the money could be used for any purpose authorized in the displaced grant programs. Thus federal oversight would be decreased and local discretion expanded. Moreover, the money would go to general purpose governments at the local level, not to the autonomous or semiautonomous local public agencies (like redevelopment agencies, housing authorities, and Model Cities' Community Development Agencies) which have been recipients of federal, urban program funding in the past.

Once again, an important political question is how the funding pie will be divided. In the administration's bill, all cities with populations of over 50,000 would receive a direct allocation determined by the formula; so would counties whose populations (besides those living in cities of over 50,000) exceeded 200,000. Three-quarters of the total appropriations would be allocated among these cities and "urban counties"; 20 percent would go to state governments (half of which would have to be distributed to metropolitan areas); and the remainder would be for discretionary spending by the secretary of HUD. The formula itself would be based on three factors: total population, the poverty population (double-weighted), and overcrowded housing.

As in the case of manpower, central cities—particularly those that have been most active in attracting federal grant programs—have been concerned about the implications of the new program for the level of funding they receive. According to data from HUD, the six largest cities would fare about as well under the Better Communities Act as they have done in the past. But those cities in the 50,000–1 million population category would experience a dramatic reduction in federal aid, from a total annual figure of $1.52 billion to $970 million. There would also be a differential geographic impact; the Northeast would be hurt while the West would gain.[19] States and counties, which heretofore have had limited experience in the area of community development, stand to gain substantially. Lobbyists for cities, together with certain senators and representatives, have been seeking to define "urban counties" in the legislation in such a way as to include some record of prior program experience and capacity, rather than simply relying on a population measure. Otherwise, city representatives say, suburban counties with few acute needs and little expertise will be getting an unduly large share of the funds. The National Association of Counties, of course, is opposed to such additional guidelines.

Community-development revenue sharing would not only alter the distribution of resources among cities, but it has implications for the power relationships within cities as well. Many officials of independent agencies, who have received substantial federal funding in the past, view with alarm the new preference for general purpose governments. Some of these people have urged that the community-development bill contain a provision allowing any general purpose government to designate a local public agency to plan and implement development policies. But some policy makers have questioned whether such wording would offer independent agencies as much protection as would other legislative provisions. A representative of the National Association of Housing and Redevelopment Officials (NAHRO) remarked, "The guidelines on spending are more important than specifying institutions. If money is earmarked for dealing with slums and blight, who can deal with slums and blight? *We* can." An HUD official, observing that independent authorities would probably capitalize on any detailed administrative requirements written into the law, said "If you get into detailed paper flow exchange—higgledy-piggledy details—the mayor will drop out. The bureaucrats in HUD and redevelopment agencies will talk back and forth about details. Mayors don't have the patience for that."

Regardless of statutory provisions, the independent agencies have some impressive resources on their side: professional expertise and years of experience in dealing with federal programs. A formal requirement of working through the mayor's office might not make much difference to these agencies, particularly if the mayor and his staff are uninterested in higgledy-piggledy details. Thus, although a change in federal policy may pose a

---

[19] See Reischauer, "The New Federalism and the Old Cities," pp. 19–22.

threat to local independent agencies—and may be used strategically by aggressive mayors who want to take over such organizations—we should be careful to keep in mind the comparative advantages that independent agencies will continue to have.

In the legislative struggle over community development, many members of Congress (particularly senators) have pressed for certain guidelines— regulating the uses of the money, setting national standards for perform- ance, providing for an application procedure, and introducing requirements for citizen participation. Administration spokesmen have continued to favor automatic distribution by formulas and much discretion regarding local use of the funds.

Behind these disagreements on the form of legislation lie more basic, philosophical differences concerning the role the federal government should play in directing the activities of local governments. Richard P. Nathan, who as an assistant director of the OMB played an important part in the shaping of special revenue sharing, has stated the case for the sup- porters of that concept:

> Are we to overhaul existing categorical grants to provide funds to elected officials on a significantly broader and less conditioned basis—giving them the freedom to make their own mistakes—as long as they operate within the framework of the Constitution and the courts as far as political processes are concerned?
>
> *Or* are we going to change the design of federal grants so that instead of defining specific program uses for federal aid funds we prescribe a series of process requirements which build in current preferred conceptions among urbanologists to take into account the interests of favored groups above and beyond the protections which the Constitution and the Courts have provided as regards their role in the political process?[20]

Those who favor Nathan's approach argue that funding by formula will allow all localities, not just those adept at grantsmanship, to share in federal resources. Moreover, they assert that the freedom to control program outcomes at the local level will produce both substantive and procedural innovations.

Critics of special revenue sharing feel that the federal government has a responsibility to ensure that poor people and minority groups, who have had difficulty in gaining access to local political systems, are not now cut off from the benefits of federal programs. They point to citizen-participation provisions of recent categorical programs as enabling weaker groups to

[20] Nathan, "Essay on Special Revenue Sharing," in Joseph D. Sneed and Steven A. Waldhorn, eds., *Approaches to Accountability in Post Categorical Programs* (Stanford, Calif.: Stanford Research Institute, 1973), p. 48.

have some voice in local decision making. And they argue that categorical programs, whatever their faults, concentrated resources on the communities that needed them and had proved that they could use them effectively, instead of spreading funds around the country. A representative of a city government, who takes a dim view of special revenue sharing, had this to say about the administration's approach, "It removes from us the opportunity to rise to incentives. . . . It reduces money for cities that have cared about poor people's problems and gives it to places who have never given a damn."

As legislative consideration of special revenue sharing goes on, there continue to be struggles both between competing jurisdictions and between competing normative views on the desirability of federal intervention in local affairs.

Transitional Measures and Some Barriers to Change

Without waiting for congressional enactment of community-development revenue sharing, the executive branch has initiated a number of administrative measures designed to increase the authority and discretion of local elected officials. Two of the most prominent of these measures have been Planned Variations and Annual Arrangements. Although certain local officials have been able to use these measures to strengthen their institutional positions, the ability to do this has been far from universal. The experience of these administrative initiatives shows the problems that local political realities can pose for federal efforts toward change, and indicates some of the outcomes and conflicts resulting from special revenue sharing.

Planned Variations

Launched by President Nixon in July 1971, this administrative initiative has sought to expand the power of 20 city governments, chosen from the 147 then participating in HUD's Model Cities program. Under Planned Variations, these city governments would gain three new powers: (1) power to spend Model Cities money citywide, according to locally determined priorities, instead of restricting spending to previously designated "model neighborhood" areas; (2) power to obtain waivers of administrative regulations in any federal program operating in conjunction with a Model Cities plan; and (3) power for the chief executive officer of the city to review and comment on all federal spending proposals in the city. In order to make the program more attractive to local recipients, the President provided the selected cities with $79 million in additional Model Cities money for fiscal 1972, and the same amount for fiscal 1973.

The centerpiece of this program was intended to be the "Chief Executive Review and Comment" provision; administration spokesmen expressed the hope that elected local leaders could use that provision to exert influence both on federal policy makers and on local recipients of federal funds. In practice, the impact of this newly granted authority has been strongly influenced by the relationships previously existing in the Planned Variations cities. For example, in Houston and Indianapolis, aggressive mayors, operating in strong-mayor governmental structures, used their new authority to force other local interest groups to support mayoral funding priorities. Mayor Louie Welch of Houston and Mayor Richard Lugar of Indianapolis also forced state and county governments to respect their preferences regarding the use of federal funds affecting their cities, and both men used the increased money and authority of the Planned Variations program to create management structures within their respective offices.[21]

But not all mayors in the Planned Variations experiment have enjoyed this kind of success. In Rochester, New York, which has a weak-mayor—strong-manager system, the new program did not result in expanded power for the mayor or other elected officials. Mayor Stephen May wanted to take charge of the city's Planned Variations effort, but he found that he lacked the legal power, staff, and budget necessary to do so.[22] And in San Jose, California, Mayor Norman Mineta was similarly disappointed by his experience with the program. As in Rochester, the city charter of San Jose accords the mayor few powers over the operating agencies. Furthermore, numerous special governmental districts in the area are outside of the city's authority. Another problem for the mayor was added when officials of Santa Clara County—a governmental unit with wide responsibilities and an extensive staff—thought that they, rather than San Jose officials, should exercise the review and comment authority. Given this local political environment, Mayor Mineta's opportunities for action were limited. The mayor's office was able to use Planned Variations money to chart the flow of federal money to city agencies; these agencies now have to get the mayor's approval before applying for a grant. But the goal of Planned Variations—that the mayor would have influence over all federal funding affecting his community—was not realized in San Jose. Mayor Mineta was unable to trace the flow of funds to special districts, independent groups, and the county, much less to review and comment on such funding.[23]

[21] See "The New Federalism: Theory, Practice, Problems," p. 56.

[22] Ibid., pp. 56–57.

[23] For an assessment of the San Jose experience, see ibid., p. 56; and Lou Cannon and David S. Broder, "Nixon's 'New Federalism': Struggle to Prove Itself," *The Washington Post*, June 17, 1973. For an overall assessment of the program, see U.S. Department of Housing and Urban Development, *Planned Variations: First Year Survey* (Community Development Evaluation Series, no. 7 (October 1972).

In some cities, local officials argued over whether the mayor, the council, or the manager was the "chief executive." There were also limits to federal cooperation with the new program. In a number of regions, federal administrators have been less than enthusiastic about extending to local officials an increased measure of control over federal funding. Furthermore, there has been some confusion over what happens after a mayor reviews and comments upon an application for federal assistance. Can the mayor use this power to veto a project? The sponsors of Planned Variations say that they hope such ultimate tests of authority can be avoided, and that the *threat* of a veto will be enough to increase a mayor's bargaining position.

In any event, the experience of the Planned Variations program shows that local political leaders vary in their capacity to utilize increased federal grants of authority. The federal government can alter a local political system by providing resources to certain actors and withdrawing them from others, but the outcome of the federal initiative will in turn be influenced by political relationships within the local system.[24]

Annual Arrangements

This administrative effort, also introduced in 1971, has had goals similar to those of Planned Variations: simplifying federal procedures and expanding the authority of local elected officials. The Annual Arrangement process was modeled after 1970 negotiations between HUD's Chicago regional office and Mayor Richard Hatcher of Gary, Indiana. In those negotiations, the city agreed to undertake a number of activities desired by HUD—including revision of building codes and provision of low-income housing—in exchange for HUD's commitment to fund a list of projects selected by Mayor Hatcher. HUD Secretary George Romney, in May of 1971, suggested that similar annual arrangements be negotiated in appropriate cities, and by March 1973, seventy-nine such arrangements had been negotiated.[25]

As envisioned by HUD officials, Annual Arrangements would aid in the transition to community-development revenue sharing by eliminating some of the procedural complexities and uncertainties of urban grant programs and by identifying the local chief executive as the prime local contact in negotiations with the federal government. Instead of dealing with numerous

[24] For further discussion of this point, see Jeffrey L. Pressman, *Federal Programs and City Politics* (Berkeley: University of California Press, 1975), especially chaps. 3 and 6.

[25] On the subject of Annual Arrangements, see "The New Federalism: Theory, Practice, Problems," pp. 61–64; and two HUD reports: *Annual Arrangements Phase I* (Community Development Evaluation Series, no. 6 (March 1972); and *Annual Arrangements: Improving Coordination of Community Development Programs* (Community Development Evaluation Series, no. 14 (May 1973).

independent, local recipients, HUD would attempt to work directly through the chief executive at city hall. HUD would agree to fund the city's priority projects, if the city would agree to undertake certain activities (low-income housing, affirmative-action employment practices, and so forth) insisted upon by HUD. The uncertainties of the project-by-project application process would be replaced by bargaining over essentials.

Although local leaders have expressed their support for the Annual Arrangements program as initially conceived, and although certain federal regional offices (notably the one in Dallas–Fort Worth) have been unusually aggressive in implementing it, the results of this effort show once again the difficulties of making structural changes in the federal system. In this case, although HUD regional and area offices made commitments to fund projects, they were not able to circumvent the established categorical processes nor to prevent HUD itself from running short of money. As one HUD official remarked, "HUD says: 'We'll be able to fund X.' But cities still have to go through the application process. And we have a shortage of dollars—we just don't have the bags of bucks. So what's the incentive [for locals] to go along?"

Furthermore, as was the case with the Planned Variations program, Annual Arrangements could not summarily abolish the realities of local political life. In Oakland, California, for example, where local officials had made a practice of shying away from involvement with federal programs—and avoiding conflict of all kinds—the Annual Arrangements program did not force a sudden change in behavior. City representatives, not wishing to disturb existing administrative relationships, were decidedly unenthusiastic about a federal request that the city create a new position of assistant city manager for community development. Local department heads were concerned about protecting their existing jurisdictions, and the city manager was not eager to upset matters by challenging the operating departments. As for the city's elected officials, HUD representatives complained that these people did not seem to take much interest in the Annual Arrangements program.[26]

It is, of course, impossible to consider the impact of Planned Variations and Annual Arrangements—or to discuss any New Federalism program—in isolation from other events that have been occurring simultaneously in the national political system. The Annual Arrangements program, for instance, was dealt a severe blow by budget cuts in community-development programs, and local officials' support for the administration's programs in general has been weakened by federal budget reductions and impound-

---

[26] See "The New Federalism: Theory, Practice, Problems," p. 63; and Pressman, *Federal Programs and City Politics*, chap. 6.

ments. Big-city mayors have charged that general revenue sharing has turned out to be a device for cutting back urban grants, despite the Nixon Administration's original promise that it would constitute "new money."

The issues of Watergate and impeachment subsequently diverted the attention of both the Nixon Administration and Congress from issues of the New Federalism. According to the "normal," incrementalist model of American politics, the executive branch and Congress share a belief that federal programs should exist in particular areas, and bargaining is confined to marginal disagreements over the shape of those programs. Both the President and Congress make concessions, because each is aware that it is dependent upon the other for the achievement of its own goals. But, in the charged atmosphere of early 1974, very little of this normal bargaining appeared to be taking place. Congressional aides in the community-development field noted that administration representatives were far less visible than usual in the legislative process. And a Senate staff member remarked: "The interesting thing about the [Nixon] administration is that it doesn't really try to bargain very much with Congress. It doesn't have a big domestic program to get through, so there's really no reason for it to make concessions to Congress." By mid-1974, the debate over the New Federalism had been submerged by even more divisive conflicts between Congress and President Nixon.

The Long-Term Political Implications

Having examined particular New Federalism programs and the political questions they have raised, let us now consider some of the long-term political implications of the New Federalism. Although the impact of the New Federalism has usually been discussed with respect to state and local governments, it is important to recognize that a shift in the structure of federal programs would have a significant impact on national governmental institutions as well. But in discussing the ways in which these program changes might be expected to alter the distribution of political resources between and within levels of government, we must once again recognize the role of existing structures and relationships in influencing the course of those changes.

Effect on the Congress

It is not surprising that members of Congress have been among the most vocal critics of proposed shifts from categorical programs to discretionary funding, because these changes pose threats to certain long-standing congressional prerogatives. First of all, to the extent that local governments gain increased authority over program design and evaluation, the role of congressional oversight is diminished. Furthermore, for individual congressmen, the switch to general purpose funding would mean fewer opportunities to identify with specific national programs and to take credit for delivering particular projects to their constituencies. Finally, increasing the resources of local officials could enable governors and mayors to improve their political positions and undertake electoral challenges to incumbent national legislators. (It is also possible that the New Federalism could enhance the position of governors vis-à-vis senators as preferred presidential timber, thus reversing the marked trend of recent years towards the Senate.)

In the struggle over New Federalism programs, as in other issue-oriented areas, Congress is not in a very strong position, for, as we have seen, it is possible for the executive branch to design administrative devices to achieve goals without legislation. And even if Congress should succeed in tying guidelines to special revenue-sharing programs, it is up to the executive branch to enforce those guidelines. A Senate staff member, in discussing this problem, provided an example of what he felt was the thwarting of legislative will: "In the New Communities program, Congress provided that the secretary [of HUD] is supposed to set aside low-income housing for those communities. Now HUD says the new communities have to provide housing for themselves." According to the aide, legislators were uncertain about what, if anything, to do next.

Effect on the Federal Bureaucracy

For members of the executive branch bureaucracy, the New Federalism would necessitate redefinitions of role. With passage of authority from the federal to local levels, it is not at all clear what the role of federal agency program managers would be. They might be called upon to provide *technical assistance* to localities, but attempts to define that term have typically been vague. For federal officials interested in carving out a social activist role for themselves, the New Federalism programs could be frustrating. As an HUD field office staffer remarked, "I came to work for the government to really have an impact on social programs. Now, with revenue sharing, who knows what I'll do? I certainly don't want to sit around here and sign checks for local governments." Actually, the position of federal field offices has always been somewhat unclear, and it will be interesting to see how the New Federalism programs will affect that position. (It could be that such offices could benefit by a general decentralization of federal authority, but their precise role under the new programs has yet to be defined.)

Effect on State Governments

Drawing upon studies of state governments which have found them to be unrepresentative, institutionally weak, of low visibility, and dominated by narrow economic interest groups,[27] critics of the New Federalism have argued that state governments are unlikely to use new resources and authority in progressive—or even efficient—ways.[28] Those who support the granting of additional authority to state governments predict that such authority will heighten public interest and participation in state policy making. Daniel J. Elazar has argued that state governments *have* become more efficient, honest, and programmatically active, and are no less able than the federal government to carry out public policies.[29]

In any event, the consequences of transferring authority from the national to the state level is of more than academic interest to groups whose strength differs from one level to the other. Labor unions, along with groups representing large cities, poor people, and minorities, have long been more effective in Washington than they have been in state capitals; these groups have been particularly concerned about the political implications of the New Federalism. It is all very well to talk about recreating Washington-centered, interest group alliances at fifty state capitals, but the problems of developing new routes of access in previously unfriendly terrain are enormous.

More attention will undoubtedly focus on the workings of state bureaucracies. In discussions about various forms of revenue sharing, it sometimes seems as if transferring authority from the national to local level is supposed to result in the disappearance of bureaucracies and the making of decisions by interested citizens at town meetings. But recent studies by Jerome T. Murphy[30] and Lawrence E. Susskind[31] of state use of federal discretionary funds show that often noted aspects of bureaucratic behavior—caution, the importance of self-maintenance, and the triumph of

[27] See, for example, McConnell, *Private Power and American Democracy*, pp. 166–195; V. O. Key, Jr., *Southern Politics in State and Nation* (New York: Knopf, 1949), and *American State Politics* (New York: Knopf, 1956); and Malcolm E. Jewell, *The State Legislature: Politics and Practice* (New York: Random House, 1963).

[28] See Michael D. Reagan, *The New Federalism* (New York: Oxford University Press, 1972), pp. 111ff.

[29] Elazar, "The New Federalism: Can the States Be Trusted?" *The Public Interest*, no. 35 (Spring 1974), pp. 89–102.

[30] Murphy, "Title V of ESEA: The Impact of Discretionary Funds on State Education Bureaucracies," *Harvard Educational Review*, vol. 43 (August 1973), pp. 362–385.

[31] Susskind, *Revenue Sharing, Block Grants, and the New Federalism—What Role for the States?* (Cambridge, Mass.: MIT–Harvard Joint Center for Urban Studies, forthcoming).

internal routines over external purposes—are very much alive at the state level.

Jurisdictional Rivalries Within the States

When public officials talk about returning money and authority to local governments, which local governments do they mean? States, counties, townships, and large and small cities have been competing with each other for shares of resources in New Federalism programs. A number of these rivalries surfaced in the administrative forerunners to special revenue sharing, as governors and county executives complained vociferously about what they felt was the promayor bias of Planned Variations and manpower program changes.[32] Rivalries among local units have been abundantly evident in the shaping of special revenue-sharing legislation. As one city lobbyist said, with reference to the "urban counties" eligible for funding through the administration's community development bill, "Of the ninety-three urban counties, six to ten are 'legit.' The rest of them are jurisdictions that have responsibilities over suburban areas. The county lobby, NACO [National Association of Counties], is trying to get as many counties in it as possible. They don't care about the function of the legislation."

The drafting and implementation of special revenue-sharing legislation seem certain to continue to stimulate rivalries among state, county, and city governments, as each attempts to secure a larger share of the pie. Obviously, the questions of distribution among subnational governments are still very much open.

Metropolitan Fragmentation and Consolidation

Because of the guarantee in the general revenue-sharing program that all general purpose local governments will receive at least 20 percent of the statewide average per capita local grant, there is a strong incentive for "dying" local governments—that is, those with decreasing functions—to stay alive. Other governments in the same county area have little incentive to swallow up the dying governments, since the money required to raise the grant up to the 20 percent floor comes from a statewide pool, not from the county area's own entitlement.[33] By creating incentives for the maintenance of existing governmental units, no matter how weak, general revenue sharing may well have adverse consequences for metropolitan consolidation.

Special revenue sharing also poses certain threats to metropolitan planning and cooperation. Federal legislation in the 1960s gave consider-

---

[32] "The New Federalism: Theory, Practice, Problems," pp. 52–54.

[33] Reischauer makes this point in his discussion of incentives. See Chapter 3 of this volume, pp. 55–56.

able stimulation to metropolitan cooperation. The Housing and Urban Development Act of 1965 set aside funds for research activities to be undertaken by regional planning agencies, and the Demonstration Cities and Metropolitan Development (Model Cities) Act of 1966 demanded that a variety of federal loans and grants be approved by metropolitan review agencies before the grants could be awarded. Under special revenue sharing, funds would pass directly through to local governments and would not be subject to a federal review power. The reduction in incentives for metropolitan cooperation is clear.

Within Cities: General Purpose Governments Versus Paragovernments

Proponents of the New Federalism have repeatedly expressed their intention to strengthen general purpose local governments and to reduce the power of *paragovernments*—a term, originated by Daniel P. Moynihan, used to describe the independent, nonprofit agencies which have served as local sponsors of federal categorical programs. Some of the best known of these independent organizations have been the Community Action agencies of OEO, Community Development agencies of Model Cities, nonprofit sponsors of the Labor Department's manpower programs, and neighborhood health centers funded by the Department of Health, Education, and Welfare (HEW). Direct provision of funds to local government will pose a threat to the continued existence of these agencies, and the new funding patterns—as we have seen—are also viewed with alarm by semiautonomous governmental bodies such as redevelopment agencies and housing authorities.

But reports of the independent agencies' demise, like those of Mark Twain's death, may be greatly exaggerated. After all, these agencies have expertise and experience in the complex technicalities of program implementation. Furthermore, it is not at all clear that mayors are eager to assume direct responsibility for controversial programs like redevelopment and public housing; having independent agencies in these policy areas provides elected officials with convenient protection from demands and criticism. This is not to deny that community-development revenue sharing would encourage mayors to exercise more control over independent agencies, or even encourage them to absorb some of their functions. But there are reasons for treating with caution any prediction of the disappearance of autonomous agencies.

Representatives of general purpose governments themselves are skeptical about the notion that they could quickly displace the independent agencies in the community-development area. They point out that past redevelopment procedures have provided for a sign-off by the city government, but that such formal procedures do not convey the realities of the administrative and political process involved. As one lobbyist for cities remarked, "They say: 'We have to run everything through city government.' As a technical matter, that's true. But actually, the redevelopment agency puts a huge amount of work into the details of the program, and they give it to the city council at the very end. [Redevelopment says] 'You've got to understand. . . . We have to have this in by June 30.' That's local approval—after a fashion." It remains to be seen how much of a change in present intracity relationships will result from changes in federal program design.

Local Government Capacity

One reason for skepticism about the ability of the New Federalism to stimulate aggressive behavior on the part of local governments is rooted in the frequent critical observations about "local government capacity." Although this term is frequently used in a technical sense—meaning weakness in personnel, planning, and management—a number of studies have deplored local governments' political inadequacies as well.[34] Effective action at the local level can be hampered by the absence of institutions with extensive formal powers, by the lack of a strong and coherent system of parties and groups, and by timid behavior on the part of elected leaders.

Federal funds can provide additional resources to local leaders, and politically adept local executives have been able to use those resources to their own advantage.[35] But, as we have seen with respect to the transitional New Federalism measures, the ability to utilize new federal resources varies among local communities and local leaders. Recent analyses of federal programs have shown that weaknesses in local political systems can create obstacles to the effective use of federal funds.[36] The New Federal-

[34] See, for example, James Q. Wilson, "The Mayors vs. The Cities," *The Public Interest*, no. 16 (Summer 1969); Daniel Pool, "Politics in the New Boston 1960–1970: A Study of Mayoral Policy Making" (Ph.D. dissertation, Brandeis University, 1974); and Jeffrey L. Pressman, "Preconditions of Mayoral Leadership," *American Political Science Review*, vol. 66 (June 1972), pp. 511–524.

[35] Perhaps the most often-described example of such a local executive is former Mayor Richard C. Lee of New Haven. For an in-depth analysis of this aggressive local leader, see Robert A. Dahl, *Who Governs?* (New Haven, Conn.: Yale University Press, 1961); and Raymond E. Wolfinger, *The Politics of Progress* (Englewood Cliffs, N.J.: Prentice-Hall, 1974).

[36] For example, see Martha Derthick, *New Towns In Town: Why a Federal Program Failed* (Washington, D.C.: The Urban Institute, 1972); and Jeffrey L. Pressman and Aaron B. Wildavsky, *Implementation: How Great Expectations in Washington Are Dashed in Oakland; Or, Why It's Amazing that Federal Programs Work at All, This Being a Saga of the Economic Development Administration as Told by Two Sympathetic Observers Who Seek to Build Morals on a Foundation of Ruined Hopes* (Berkeley: University of California Press, 1973).

ism cannot abolish the problem of local political capacity. Because of the tendency on the part of many local officials to try to avoid conflict and minimize their own responsibilities, we should not be too ready to assume that such officials will welcome the disappearance of federal guidelines. After all, guidelines provide an excellent scapegoat for local leaders who wish to avoid being held responsible for what may be an unpopular act. Even aggressive mayors can benefit from the opportunity to say that "the feds are making me do it" in backing a controversial cause. In an interview, one HUD official commented that he was "amazed to discover that mayors actually *want* guidelines." Another HUD representative has observed, "Mayors do want at least some national objectives [in HUD legislation] so that they have a rationale they can point to for their allocation of resources." As Thomas C. Schelling has pointed out, the power of a bargainer can be enhanced by a firm commitment to a prior position and a consequent inability to meet others' demands.[37] Federal guidelines, though much maligned in the rhetoric of local officials, do provide those same officials with an excuse to refuse to make concessions to other local actors. Thus, binding guidelines can actually provide leaders with an increased measure of freedom in the local arena. No wonder the abolition of guidelines is less than unanimously supported by mayors.

Questions of Redistribution and Race

Much of the opposition to the change from categorical programs to various forms of revenue sharing comes from spokesmen for racial minorities and from those who support redistributive social programs. A range of Great Society categorical measures—the poverty program, Model Cities, Concentrated Employment Program, and others—were directed toward the particular needs of poor people and minorities. Based on past performance and the characteristics of local political systems, it is doubtful that states and local governments will make similar commitments of resources to politically powerless groups.

Daniel P. Moynihan, who is now regarded as a prime exponent of the New Federalism, wrote in 1967 about some of the race-related reasons for deciding policy questions at the national level:

> The necessity for concentrating decision making at the national level will be enhanced if current trends in racial concentration persist. Between 1960 and 1966, the number of children under age fourteen in metropolitan areas increased by 3.3 million. . . . The average annual rate of increase of nonwhite children (2.4 per cent) was three times the rate for white children. Ninety-five

[37] Schelling, *A Strategy of Conflict* (New York: Oxford University Press, 1960), p. 19.

per cent of the nonwhite increase was in central cities. . . . According to one estimate, by 1970 Negroes will constitute 40 per cent or more of the population in fourteen of the nation's major cities. . . . In southern communities accustomed to taking collective measures to prevent Negro accession to power, there may be movements toward metropolitan governments in order to maintain Negroes in a minority voting status; but, in general, continued and possibly heightened racial tension is likely to inhibit greatly the development of true metropolitan governments. *A fortiori* the resolution of conflict between central cities and suburbs (which will increasingly take on "urban" qualities of their own) will have to occur at the federal level, save for the few states with sufficient political and fiscal resources to handle such matters at the level of state government.[38]

Arguments over national versus local control in this country have long been intertwined with questions of race, and the problems to which Moynihan pointed have not disappeared. The undistinguished record of local governments in this area has led minority group spokesmen to be less than optimistic about their performance in the future.

Looking Ahead: A Return to Categoricals?

Even if there are extensive shifts from national categorical programs to local discretionary ones, there are reasons for thinking that the pendulum will eventually swing back in the other direction. Reischauer argues that pressures would eventually build for federal supervision of string-free block grants, including guidelines on spending levels and practices.[39] And Herbert Kaufman has suggested that cycles of centralization and decentralization will alternate with each other. Just as centralized policy making has led to calls for increased responsiveness to local needs, so decentralization would stimulate calls for equality between units and for concerted national action.[40]

Regardless of the fate of specific New Federalism programs, the struggle between proponents of centralization and decentralization, national power and local power, will continue to be a central one in the public arena. Far from being a technical matter of administration, the design of federal programs has important implications for the distribution of scarce resources among jurisdictions, institutions, and groups. The intensity of the political struggles over the New Federalism indicates that many people understand that.

[38] Moynihan, "The Relationship of Federal to Local Authorities," *Daedalus,* vol. 96 (Summer 1967), p. 802.

[39] Reischauer, "The New Federalism and the Old Cities," p. 34.

[40] Kaufman, "Administrative Decentralization and Political Power," *Public Administrative Review,* vol. 29 (January–February 1969).

3 General Revenue Sharing— The Program's Incentives

ROBERT D. REISCHAUER*

Categorical grants have been severely criticized in recent years because the incentives, restrictions, and regulations that characterize such inter-governmental transfers are thought to have altered the behavior of re-cipient governments in socially undesirable ways. It has been argued that the availability of federal aid for certain narrow, functional areas has led to the provision of these services at the expense of others that have much greater priority from the local perspective. Matching requirements al-legedly have distorted spending patterns and lured some jurisdictions into fiscal overcommitment. Certain federal restrictions are thought to have fostered inefficiencies. For example, regulations limiting mass transit aid to capital projects has led to the premature purchase of new buses when more effective maintenance of the existing stock was more efficient.[1] Finally, many think that the methods used to allocate categorical grants

*Research Associate, The Brookings Institution.

AUTHOR'S NOTE: Without implicating them in any way for the shortcomings of this paper, I am indebted to Allen Manvel and Richard Nathan for many constructive sug-gestions and for generously allowing me to use data developed for the volume, *Monitor-ing Revenue Sharing* by Richard Nathan, Allen Manvel, and Susannah Calkins (Wash-ington, D.C.: Brookings Institution, 1975). The views expressed in this paper are my own, and do not necessarily reflect those of the rest of the staff, the officers or the trustees of the Brookings Institution.

[1] See William B. Tye, "The Capital Grant as a Subsidy Device: The Case of Urban Mass Transportation," in *The Economics of the Federal Subsidy Program*, pt. 6, *Trans-portation Studies*, Joint Economic Committee (Feb. 26, 1973). A similar situation in water resources is described in James C. Loughlin, "Federal Reimbursement Policy for Water Resources and Other Programs," *Public Finance Quarterly*, vol. 2, no. 1 (January 1974), pp. 86–106.

have encouraged governments to waste considerable sums just in obtaining federal aid; that is, they have invested in grantsmanship.

In part, general revenue sharing represented a response to these criticisms. According to the rhetoric, this program is free of the distorting elements that are the hallmarks of categorical grants; as President Nixon so eloquently put it when signing the State and Local Fiscal Assistance Act of 1972 which established the program, "When we say no strings, we mean no strings."[2] This view that the revenue-sharing program is a completely neutral grant seems plausible considering first, that the money is automatically distributed according to a formula to all those jurisdictions which meet certain objective eligibility criteria and, second, that the recipients are left free to spend the money on a wide range of public services.

In reality, however, the general revenue-sharing program is not "string-free"; it does contain a number of restrictions and incentives which in the long run could significantly affect the behavior of state and local governments. While some of these were clearly intended by the Congress and others were at least dimly perceived by the program's architects, there are a few that may turn out to be something of a surprise. This paper describes these incentives and strings, examining the extent to which they may influence the behavior of recipient governments. Since these incentives arise largely from the workings of the distribution formula, a fairly complete description of that formula—how it works and who gets what from it—is included in the next section, and this discussion is followed by an analysis of the various incentives.

The Distribution Formula: Who Gets What and Why?

During the five-year period from January 1972 to December 1976, the revenue-sharing program will distribute $30.2 billion among the states, general purpose local governments, and Alaskan native villages and Indian tribes. The grants received by governments for 1972 ranged from New York City's entitlement of $214 million to the myriad of checks that just exceeded the minimum allowable grant of $200.[3] On a per capita basis the

---

[2] "Statement by the President upon Signing the Bill Providing State and Local Fiscal Assistance," Oct. 20, 1972, *Weekly Compilation of Presidential Documents* (Washington, D.C.: Office of the Federal Registrar, Oct. 23, 1972), vol. 8, no. 43, p. 1535.

[3] If Indian tribes not subject to the $200 limit are included, the smallest annual entitlement was the $26 given to the one-member Cortina Rancheria Indian tribe. All of the figures in this paper refer to the amounts received for the program's first full year, calendar year 1972. The amounts received by any particular government vary from year to year, both because some of the data upon which the distribution is based are updated periodically and because the amount to be distributed increases gradually over the program's five-year life.

TABLE 1. State, County, Municipal, and Township Governments by Size, 1972[a]

Size	No.	Percentage of total
Total	**38,602**	**100**
Over 100,000	542	1.4
50,000–99,999	624	1.6
10,000–49,999	3,956	10.3
2,500–9,999	6,321	16.4
1,000–2,499	7,225	18.7
Under 1,000	19,934	51.6

Source: Derived from data calculated by Allen Manvel from *1972 Census Governments*, vol. 1, *Governmental Organization.*

[a] These figures do not reflect the size distribution of recipient Indian tribes and Alaskan native villages but do include roughly 1,200 townships and municipalities that for various reasons do not receive revenue-sharing checks. The vast majority of these units fall into the size categories under 2,500.

entitlements ranged from the $38.55 received by nine municipalities in Mississippi to the $0.004 provided Niles township, Illinois. All told, some 38,000 governmental units received checks. These jurisdictions range in size from the state of California with a population of over 19 million to a number of tiny towns and municipalities such as Kingsbury, Maine, with fewer than ten inhabitants. Fewer than one-third of the recipient jurisdictions have populations exceeding 2,500, and roughly half of the recipient jurisdictions have fewer than 1,000 inhabitants (Table 1).

Interstate Allocations

The division of the money among the recipient governments is determined by the revenue-sharing formula, or, more correctly, formulas. First, the amount appropriated for any entitlement period is divided among state areas. (The term *state area* refers to the state government and the local general purpose governments and native villages or Indian tribes within the state.) Each area is allotted a grant equal to the larger of the amounts due it under the distribution formula first passed by the Senate and the distribution formula initially approved by the House of Representatives.[4]

[4] Since the sum of these allotments would be larger than the amount appropriated for any entitlement period, the allotments are all reduced by a percentage equal to

$$1 - \frac{A_T}{\sum_{i=1}^{51} G_{iT}}$$

where

A_T = the amount appropriated for the entitlement period (T) and

G_{iT} = the larger of the amount due state area (i) in period T under the Senate or House formula.

This reduction meant that thirteen states received less, or were worse off, in 1972 than they would have been if either of the formulas alone had prevailed.

The Senate formula (1) distributes revenue-sharing money on the basis of population, general tax effort, and relative per capita income combined in a multiplicative fashion[5]:

$$(1) \qquad G_{jT} = A_T \left[\frac{POP_j \left(\dfrac{S + LTAX_j}{AGINC_j} \right) \left(\dfrac{PCY_{US}}{PCY_j} \right)}{\displaystyle\sum_{all\ i} POP_i \left(\dfrac{S + LTAX_i}{AGINC_i} \right) \left(\dfrac{PCY_{US}}{PCY_i} \right)} \right]$$

where

G_{jT} = revenue-sharing grant to state j in entitlement period T

A_T = the total amount appropriated for entitlement period T

POP_i = population of state i

$S + LTAX_i$ = net taxes collected by state i and all its local governments

$AGINC_i$ = aggregate personal income of state i (Department of Commerce figures used for National Income Accounts purposes)

PCY_i = per capita income of state i (Bureau of the Census sources).

On a per capita basis, state area allotments under the Senate formula for the first two entitlement periods (calendar year 1972) varied from 167 percent of the national average in Mississippi to 75 percent of the average in Connecticut and 72 percent in the District of Columbia (see Table 2). This variation reflects differences in each state's per capita income relative to the national average and in the fraction of the state's aggregate income absorbed by all state and local taxes (general tax effort). The values of these variables are provided in the first two columns of Table 3.

The House formula (formula 2) allocates revenue-sharing money on the basis of urbanized population and state income tax collections as well as those factors found in the Senate's formula.

$$(2) \quad G_{jT} = A_T \left[0.2201 \frac{POP_j}{POP_{US}} + 0.2201 \frac{URBPOP_j}{URBPOP_{US}} \right.$$

$$+ 0.2201 \frac{POP_j \left(\dfrac{PCY_{US}}{PCY_j} \right)}{\displaystyle\sum_{i=1}^{51} POP_i \left(\dfrac{PCY_{US}}{PCY_i} \right)} + 0.1698 \frac{\dfrac{(S + LTAX_j)}{AGINC_j}(S + LTAX_j)}{\displaystyle\sum_{i=1}^{51} \dfrac{(S + LTAX_i)}{AGINC_i}(S + LTAX_i)}$$

$$\left. + 0.1698 \frac{YTAX_j}{\displaystyle\sum_{i=1}^{51} YTAX_i} \right]$$

[5] Alaska and Hawaii receive a supplement the size of which is based on the wage adjustment given to federal civilian employees in those states to compensate for higher living cost. In 1972, this supplement was 15 percent in Hawaii and 25 percent in Alaska.

TABLE 2. Per Capita Distribution for Entitlement Period 1 Under the Senate (S) and House (H) Formulas, and the State and Local Fiscal Assistance Act of 1972, Relative to the U.S. Average (U.S. Average = 100)

State	Operative formula	Senate formula	House formula	Ratio Senate/ House	State and Local Fiscal Assistance Act
Alabama	S	110	81	135	101
Alaska	S	92	86	107	84
Arizona	S	118	101	118	109
Arkansas	S	120	77	156	109
California	H	92	117	85	108
Colorado	H	101	103	98	95
Connecticut	H	75	93	81	85
Delaware	H	94	122	77	112
District of Columbia	H	72	132	55	121
Florida	S	90	85	106	83
Georgia	S	100	86	116	92
Hawaii	H	126	129	98	118
Idaho	S	125	85	147	114
Illinois	H	86	103	83	95
Indiana	S	92	86	107	84
Iowa	S	112	91	123	102
Kansas	S	98	80	121	89
Kentucky	S	113	85	132	104
Louisiana	S	141	89	158	129
Maine	S	131	76	171	120
Maryland	H	90	114	79	105
Massachusetts	H	97	122	80	111
Michigan	H	95	106	90	97
Minnesota	H	114	117	97	107
Mississippi	S	167	79	211	153
Missouri	H	87	88	99	81
Montana	S	124	93	133	113
Nebraska	S	109	87	126	100
Nevada	H	95	99	96	90
New Hampshire	S	94	72	130	86
New Jersey	H	78	97	80	89
New Mexico	S	136	85	160	124
New York	H	105	135	78	124
North Carolina	S	112	86	131	103
North Dakota	S	150	76	197	138
Ohio	H	76	84	90	77
Oklahoma	S	96	79	122	88
Oregon	H	96	106	90	97
Pennsylvania	H	97	99	99	90
Rhode Island	H	101	107	95	98

TABLE 2. (Continued)

State	Operative formula	Senate formula	House formula	Ratio Senate/ House	State and Local Fiscal Assistance Act
South Carolina	S	116	82	142	107
South Dakota	S	151	77	198	139
Tennessee	S	105	78	136	97
Texas	S	93	85	108	85
Utah	S	121	104	116	110
Vermont	S	139	95	146	127
Virginia	H	91	96	95	88
Washington	S	96	87	110	88
West Virginia	S	125	80	155	114
Wisconsin	S	126	118	107	116
Wyoming	S	126	72	176	115

Source: Calculated from data published in Staff of the Joint Committee on Internal Revenue Taxation, *General Explanation of the State and Local Fiscal Assistance Act and the Federal–State Tax Collection Act of 1972* (Feb. 12, 1973), p. 26.

where

$URBPOP_i$ = urbanized population of state i

$YTAX_i$ = net state income tax collections of state i (subject to constraints).

The factors which are expressed as ratios, for example, relative income and general tax effort, are weighted by population and state and local tax collections, respectively. Such a weighting procedure is necessitated by the formula's additive nature. If it were not employed, two states with similar relative incomes but of very different populations, such as Nevada and New York, would be accorded roughly equivalent, aggregate amounts of money by the relative income factors; on a per capita basis, however, this would mean that Nevada would receive more than forty times the allotment of New York.

The values of the factors used in the House formula for the first entitlement period are provided in columns 3 through 7 of Table 3. Since Alaska, Wyoming, and Vermont do not contain a city large enough to be considered a central city, they have no urbanized population (column 4). While a number of states did not collect income taxes, the income tax term (column 7) is never zero because the act permits states with low or nonexistent state income taxes to substitute an amount equivalent to 6.67 percent of the federal income tax liability of their residents into the formula in lieu of the state income tax figures. If the House formula had been

TABLE 3. Factors Influencing the Per Capita Revenue-Sharing Allotments for Entitlement Period 1

State	Senate formula				House formula		
	$\dfrac{PCY_{US}}{PCY_i}$	$\dfrac{S + LTAX_i}{AGINC_i}$	$\dfrac{POP_i}{POP_{US}}$ [a]	$\dfrac{URBPOP_i}{URBPOP_{US}}$ [a]	$POP_i \dfrac{\frac{PCY_{US}}{PCY_i}}{\sum POP_i \frac{PCY_{US}}{PCY_i}}$ [a]	$\dfrac{\frac{(S + LTAX_i)^2}{AGINC_i}}{\sum \frac{(S + LTAX_i)^2}{AGINC_i}}$	$\dfrac{YTAX_i}{\sum YTAX_i}$ [a]
Alabama	1.35	9.76	1.69	1.08	2.23	0.81	0.83
Alaska	0.84	10.43	0.15	0.0	0.12	0.13	0.28
Arizona	1.06	13.32	0.87	0.98	0.91	0.99	0.63
Arkansas	1.46	9.73	0.95	0.30	1.35	0.44	0.44
California	0.86	13.73	9.82	13.63	8.30	14.56	12.39
Colorado	1.00	12.06	1.09	1.20	1.07	1.09	1.21
Connecticut	0.80	11.11	1.49	1.78	1.17	1.59	0.83
Delaware	0.96	11.70	0.27	0.30	0.25	0.28	0.61
District of Columbia	0.81	10.69	0.37	0.64	0.30	0.40	0.81
Florida	1.02	10.58	3.34	3.49	3.34	2.43	0.12
Georgia	1.18	10.09	2.26	1.59	2.61	1.36	1.78
Hawaii	0.92	14.05	0.38	0.37	0.34	0.59	0.87
Idaho	1.18	12.64	0.35	0.72	0.41	0.32	0.37
Illinois	0.89	11.47	5.47	6.64	4.78	5.73	5.69
Indiana	1.02	10.76	2.56	2.02	2.54	1.98	1.69
Iowa	1.08	12.34	1.39	0.71	1.47	1.38	1.44
Kansas	1.06	10.93	1.11	0.66	1.15	0.89	0.57
Kentucky	1.29	10.48	1.58	0.95	2.0	0.94	1.16
Louisiana	1.34	12.55	1.79	1.44	2.35	1.52	0.69
Maine	1.22	12.75	0.49	0.15	0.59	0.46	0.17
Maryland	0.89	12.11	1.93	2.19	1.68	2.14	3.38
Massachusetts	0.92	12.71	2.80	3.66	2.51	3.49	4.98
Michigan	0.93	12.24	4.37	4.78	3.97	4.70	5.58

Minnesota	1.03	13.25	1.87	1.61	1.88	2.22	3.79
Mississippi	1.62	12.30	1.09	0.27	1.73	0.75	0.33
Missouri	1.06	9.87	2.30	2.18	2.38	1.50	1.58
Montana	1.16	12.74	0.34	0.12	0.39	0.33	0.44
Nebraska	1.12	11.72	0.73	0.50	0.80	0.67	0.45
Nevada	0.88	12.96	0.24	0.28	0.21	0.33	0.12
New Hampshire	1.04	10.75	0.36	0.15	0.37	0.27	0.14
New Jersey	0.81	11.00	3.53	5.13	2.78	3.48	1.73
New Mexico	1.28	12.65	0.50	0.25	0.63	0.44	0.28
New York	0.86	14.54	8.97	12.05	7.59	16.00	18.56
North Carolina	1.26	10.60	2.50	1.02	3.09	1.59	2.46
North Dakota	1.26	14.19	0.30	0.45	0.38	0.32	0.11
Ohio	0.97	9.25	5.24	5.61	5.00	3.15	2.29
Oklahoma	1.16	9.93	1.26	0.89	1.43	0.73	0.52
Oregon	0.99	11.55	1.03	0.83	1.00	0.90	1.85
Pennsylvania	1.02	11.39	5.80	5.84	5.78	5.22	5.97
Rhode Island	1.00	12.05	0.47	0.63	0.46	0.47	0.45
South Carolina	1.35	10.26	1.27	0.55	1.69	0.70	0.91
South Dakota	1.31	13.84	0.33	0.64	0.42	0.35	0.75
Tennessee	1.27	9.93	1.93	1.26	2.39	1.04	0.56
Texas	1.12	9.90	5.51	5.84	6.02	3.38	1.86
Utah	1.16	12.47	0.52	0.62	0.59	0.46	0.50
Vermont	1.13	14.68	0.22	0.0	0.24	0.29	0.34
Virginia	1.04	10.43	2.29	2.02	2.33	1.59	2.73
Washington	0.93	12.28	1.68	1.58	1.53	1.79	0.63
West Virginia	1.34	11.12	0.86	0.29	1.12	0.57	0.55
Wisconsin	1.03	14.64	2.17	1.74	2.19	3.05	4.17
Wyoming	1.08	13.90	0.16	0.0	0.17	0.20	0.57

Source: Department of the Treasury, Office of Revenue Sharing, "Data Used for Interstate Allocation" (no date).

ª Multiplied by 100.

TABLE 4. State Income Tax Collections as a Fraction of Federal Income Tax Liability Attributed to State Residents

State	Percentage	State	Percentage
Alabama	14.2	Montana	29.2
Alaska	27.2	Nebraska	12.7
Arizona	13.8	Nevada	0.0ª
Arkansas	13.8	New Hampshire	2.0ª
California	20.7	New Jersey	0.7ª
Colorado	20.7	New Mexico	14.2
Connecticut	3.1ª	New York	29.5
Delaware	30.6	North Carolina	23.7
District of Columbia	31.5	North Dakota	10.0
Florida	0.0ª	Ohio	7.3
Georgia	17.7	Oklahoma	9.5
Hawaii	34.3	Oregon	34.2
Idaho	27.0	Pennsylvania	17.2
Illinois	14.6	Rhode Island	18.0
Indiana	12.0	South Carolina	24.6
Iowa	21.1	South Dakota	0.0ª
Kansas	10.3	Tennessee	1.1ª
Kentucky	18.4	Texas	0.0ª
Louisiana	10.2	Utah	23.9
Maine	8.5	Vermont	33.3
Maryland	25.6	Virginia	21.7
Massachusetts	27.3	Washington	0.0ª
Michigan	20.7	West Virginia	15.2
Minnesota	41.5ᵇ	Wisconsin	37.1
Mississippi	9.9	Wyoming	0.0ª
Missouri	12.9		

Source: Department of the Treasury, Office of Revenue Sharing, "Data Used for Interstate Allocation" (no date).
ª Affected by floor.
ᵇ Affected by ceiling.

mandated, ten states would have benefited by this floor (Connecticut, Florida, Nevada, New Hampshire, New Jersey, South Dakota, Tennessee, Texas, Washington, and Wyoming). As it is, seven of these received more under the Senate formula, and therefore received their allotment under a method which does not contain an income tax factor. At the other extreme, state income tax collections in excess of 40 percent of the federal income tax liability of any state's residents are excluded from the calculation. Minnesota alone was affected by this provision in 1972, but Wisconsin, Oregon, Hawaii, and Vermont were close to the limit (Table 4).

The variation in relative per capita allotments provided by the House formula is from 72 percent of the national average in Wyoming to 135 percent in New York (see Table 2, column 2). The way in which various factors influence the per capita distribution can be seen by contrasting the fraction of the U.S. population residing in a state (see Table 3, column 3), with the values of the other distributional factors (columns 4 through 7). In New York, for example, only the relative income factor acted to pull the state's per capita grant below the average; the other factors were all substantially larger than the share of the nation's population living in New York (8.97) and therefore acted to boost that state's per capita grant above the average.

All told, thirty-one states received their revenue-sharing allotments based on the Senate formula, while nineteen states and the District of Columbia received their revenue-sharing grants through the House formula.[6] Although both formulas contain population, general tax effort, and relative income terms there is no systematic relationship between the relative amount accorded a state under the Senate formula and that provided by the House formula.[7] Four of the five states receiving the most in per capita terms under the House formula receive less than the median state under the Senate formula. Three of the five states receiving the largest per capita grants under the Senate formula fall into the bottom fifth of the states under the House formula. In fact, there seems to be little rhyme or reason to the variation in distribution that result from the political compromise of using two formulas. Certainly, the simple correlations between per capita revenue-sharing allotments and any of the crude measures of need are not overwhelming.[8]

Intrastate Allocations

One-third of each state's entitlement is given to the state government. The remaining two-thirds is allocated among the local general purpose governments and Indian tribes of the states. This money is first apportioned

---

[6] Since many of the data elements in the interstate distribution formulas are updated annually, the particular formula by which a specific state receives its grant can shift from entitlement period to period. For example, by the fifth entitlement period, Massachusetts, Minnesota, and Pennsylvania had shifted from the House to the Senate formula and Alaska had shifted in the other direction.

[7] The simple correlation between the per capita grant under the Senate and House formulas is −0.281.

[8] Correlations of per capita revenue sharing with:

Per capita income (1971)	−0.26
ACIR's index of relative fiscal capacity (1971)	−0.16
FBI crime rate (1971)	−0.09
Urban population (percentage)	−0.43
Population in cities over 100,000 (percentage)	−0.12

among the county areas of each state on the basis of Formula 3, which is very similar to the Senate's interstate distributional formula.

$$(3) \qquad G_{jT} = \tfrac{2}{3}G_{sT}\left[\frac{POP_j \dfrac{PCY_s}{PCY_j}\dfrac{LTAX_j}{AGINC_j}}{\sum_{\text{all } i} POP_i \dfrac{PCY_s}{PCY_i}\dfrac{LTAX_i}{AGINC_i}}\right]$$

where

G_{sT} = the grant to state area s in entitlement period T

G_{jT} = allocation to county area j in entitlement period T

POP_j = population of county area j

$PCY_{s,j}$ = per capita income in state s or county j

$LTAX_j$ = net *nonschool* taxes of local *general purpose* governments in county area j

$AGINC_j$ = aggregate personal income in county j (data obtained from the Bureau of the Census)

The major difference between the two lies in the definition of the terms used to compute the tax effort factor; taxes ($LTAX$) are restricted to those collected by general purpose governments (county governments, townships, and municipalities) less their levies used to finance schools and the aggregate income data is taken from the Bureau of the Census sources.[9]

A portion of each county area's allotment equal to the fraction of the county's total population belonging to Indian tribes (native villages in Alaska) is set aside to be divided among these tribes on the basis of the size of their membership. Only 172 of the 3,044 county areas in the United States are affected by this provision, and in only six states does the share provided to Indian tribes exceed 1 percent of the state area's total entitlement (Table 5). After the Indians' share, if any, has been taken out of the county area's entitlement, the remainder is apportioned into three separate pots of money—one for the county government, one to be shared by the municipal governments in the county area, and one to be divided among the county's townships. (Only twenty-one states have both municipalities and townships as defined by the Bureau of the Census.) The fraction of the county area's entitlement going into each pot is proportional to the share

---

[9] This latter fact means that the intrastate distributional formula can be expressed as a function of per capita income and tax collections

$$G_{jT} = \tfrac{2}{3}G_{sT}\left[\frac{\dfrac{LTAX_j}{PCY_j^2}}{\sum_{\text{all } i} \dfrac{LTAX_i}{PCY_i^2}}\right]$$

because the PCY_s terms cancel out and $AGINC_j = (PCY_j)(POP_j)$.

TABLE 5. Percentage Distribution of Revenue-Sharing Money by Type of Government, for 1972[a]

State	State govern- ments (%)	Local govern- ments (%)	Indian tribes (%)	State	State govern- ments (%)	Local govern- ments (%)	Indian tribes (%)
Alabama	33.33	66.67	0	Missouri	33.36	66.64	0
Alaska	33.40	64.31	2.29	Montana	33.33	64.33	2.34
Arizona	33.33	63.58	3.09	Nebraska	33.33	66.53	0.14
Arkansas	35.96	64.04	0	Nevada	33.33	66.14	0.53
California	33.33	66.65	0.02	New Hampshire	33.33	66.67	0
Colorado	33.33	66.60	0.07	New Jersey	33.33	66.67	0
Connecticut	33.33	66.67	0	New Mexico	34.81	60.93	4.26
Delaware	40.15	59.85	0	New York	33.33	66.64	0.03
District of				North Carolina	33.33	66.59	0.08
Columbia	0	100.0	0	North Dakota	33.33	65.25	1.42
Florida	33.33	66.66	0.01	Ohio	33.33	66.67	0
Georgia	33.35	66.65	0	Oklahoma	33.33	66.13	0.54
Hawaii	33.34	66.66	0	Oregon	33.33	66.56	0.11
Idaho	33.33	66.31	0.36	Pennsylvania	33.33	66.67	0[b]
Illinois	33.33	66.67	0	Rhode Island	33.33	66.67	0
Indiana	33.33	66.67	0	South Carolina	34.23	65.77	0
Iowa	33.33	66.65	0.01	South Dakota	33.33	64.32	2.35
Kansas	33.33	66.65	0.01	Tennessee	33.33	66.67	0
Kentucky	41.38	58.62	0	Texas	33.37	66.62	0
Lousiana	33.90	66.09	0	Utah	33.34	66.15	0.51
Maine	33.33	66.53	0.14	Vermont	33.36	66.64	0
Maryland	33.33	66.67	0	Virginia	33.33	66.67	0[b]
Massachusetts	33.33	66.67	0	Washington	33.33	66.42	0.25
Michigan	33.33	66.66	0.01	West Virginia	45.24	54.76	0
Minnesota	33.33	66.47	0.20	Wisconsin	33.33	66.57	0.1
Mississippi	33.89	66.05	0.06	Wyoming	33.33	65.70	0.97

[a] Figures may not sum to 100 because of rounding.

[b] Positive but rounds to zero.

of the county area's aggregate nonschool taxes ($LTAX_j$) collected by each of these governmental types.[10] The allotments for the municipalities and

[10]

$$LTAX_j = LTAX_c + \sum_{\text{all } m} LTAX_m + \sum_{\text{all } t} LTAX_t$$

where the subscript c refers to the county government, m to a municipality, and t to a township. The shares for the county government, the municipalities, and the townships are therefore

$$G_{iT}^*\left[\frac{LTAX_c}{LTAX_j}\right]; \qquad G_{iT}^*\left[\frac{\Sigma LTAX_m}{LTAX_j}\right]; \qquad G_{iT}^*\left[\frac{\Sigma LTAX_t}{LTAX_j}\right]$$

where G_{iT}^* is the grant to a county area less the amount set aside for Indian tribes.

for the townships are apportioned among the qualifying jurisdictions on the basis of a formula which is identical to the intercounty distribution formula (see Formula 3).

The formulas for distributing revenue-sharing money among units of local government are not allowed to operate in an unconstrained fashion. There are four limits. First, no county area or unit of local government can receive more than 145 percent of the statewide average per capita amount destined for local governments. Second, with the exception of county governments, no unit or county area can receive less than 20 percent of this statewide average.[11] Third, no local government may receive an amount in excess of one-half of its net nonschool taxes plus its intergovernmental receipts.[12] Finally, if application of the formula results in a town or municipality receiving an entitlement of less than $200, the entitlement is transferred to the county government.

Complex procedures are used to make the adjustments necessitated by these limits.[13] At the county area level, money produced by the 145 percent ceiling is redistributed proportionately among the unconstrained county areas. Similarly, in the few instances where resources are needed to boost some county areas up to the 20 percent floor, the amount going to unconstrained areas is reduced proportionately to raise the amount needed for this. The money produced by imposing the 145 percent ceiling on townships and municipalities is used to boost other such governments up to the 20 percent floor. In cases where the funds generated by the 145 percent

---

[11] These limits are

$$\frac{1.45(\frac{2}{3})G_{sT}}{POP_s} \quad \text{and} \quad \frac{0.2(\frac{2}{3})G_{sT}}{POP_s}$$

where G_{sT} is the entitlement to state area s in period T.

[12] Excepting general revenue sharing, all intergovernmental transfers to general purpose local governments from state and federal sources are included, even those destined to be used for education. The order in which the constraints are applied and the interpretation of these limits has a profound effect on the distribution. For a discussion of this, see Otto G. Stolz, "Revenue Sharing—New American Revolution or Trojan Horse," *Minnesota Law Review*, vol. 58, no. 1 (November 1973), pp. 45–51.

[13] This is not a description of the computer program the Treasury Department uses to allocate shared revenue, but rather of the effect of that program. To avoid certain legal problems and for computational ease, the Treasury program, through a process of trial and error, settles on a hypothetical amount to be distributed to local governments which differs from the actual amount by roughly the difference between the excess money produced by applying the 145 percent limit to the municipalities and townships and the sum needed to bring all such governments up to the 20 percent floor. The program then takes this hypothetical amount, distributes it, and makes the necessary adjustment for townships and municipalities required by the 20 percent floor and 145 percent ceiling without altering the amounts calculated for other jurisdictions.

limit are insufficient to cover the resources needed to bring all townships and municipalities up to the 20 percent floor, a wholesale, proportioned grant reduction will take place. This cuts into the amounts received by unconstrained local governments even in county areas where no unit is affected by a limit. The only unconstrained local governments that are spared are those located in the county areas which were themselves constrained. When the 145 percent limit produces more revenue than what is needed to pull jurisdictions up to the 20 percent floor, the surplus is prorated among all units of local government which themselves are not affected by a restriction and which are not in a constrained county area.[14] When a township or municipality is hit by the restriction limiting a unit's grant to 50 percent of its taxes and transfers, the excess amounts entitled to it through the formula are given to the county government. When the county government bumps against this 50 percent ceiling, excess funds are transferred to the state government. As a result of this, a number of state governments—including West Virginia, Kentucky, and Delaware—received well over one-third of the total revenue-sharing allotments provided their state areas in 1972 (see Table 5). The maximum of 50 percent of adjusted taxes plus intergovernmental receipts takes precedence over the 20 percent of the average per capita statewide distributions. Therefore, 1,569 of the townships and municipalities that get shared revenue receive less than the 20 percent floor.[15]

The impact that the limits and constraints have on the distribution of revenue-sharing funds can be substantial.[16] All told, some $350 million, or 6.6 percent, of the first year's entitlement was redistributed through the limits.[17] Only seventy-four of the 38,000 recipients have allocations which are unaffected either directly or indirectly by these barriers; 13,196

[14] In entitlement period 3, the 145 percent surplus was larger than the money needed by the 20 percent floor in thirty-four states and the reverse was true in fourteen states. In Hawaii and Nevada no unit of local government was affected by a constraint.

[15] The Treasury computer program raises the allocations of such units to the 50 percent ceiling, thus minimizing the amount transferred to county governments from unconstrained units in other counties.

[16] A more detailed analysis of these and other formula issues can be found in Richard Nathan, Allen Manvel, and Susannah Calkins, *Monitoring Revenue Sharing* (Washington, D.C.: The Brookings Institution, 1974), chaps. 4 and 7. Many of the specific examples and numbers in these paragraphs come from data and programs developed for that volume.

[17] This figure is the sum of the absolute value of the differences between the amount given each unit under constrained formula and the amount each would receive if the formula were unconstrained. It therefore counts each dollar twice. Looking just at the allotments to local governments, the comparable figures are $345 million and 10 percent, respectively. In the states of West Virginia and Missouri, 25 and 19 percent, respectively, of the total local entitlement was redistributed by the constraints in 1972.

TABLE 6. Number of Recipient Units Directly Affected by Limits Under the Actual Allocation (A) and Number That Would Be in Violation of a Limit Under an Unconstrained Formula (U), 1972[a]

Limits	County areas	Counties	Municipalities and townships
At or over 145 percent limit	535	n.a.	1,239 (A)
			1,450 (U)
At or under 20 percent limit	5	n.a.	7,274 (A)
			8,837 (U)
At or over 50 percent of adjusted taxes and intergovernmental revenue	n.a.	173 (A)	2,867 (A)
		161 (U)	1,537 (U)
Under $200[b]	n.a.	n.a.	1,103 (A)
			1,046 (U)

Source: Data generated from computer programs and data tapes developed for the Monitoring Revenue Sharing Project, The Brookings Institution, Washington, D.C.

Abbreviation: n.a., not applicable.

[a] The differences arise from the redistribution caused by applying the limits. For example, if the 145 percent limit did not reduce the shared revenue available to some county areas more of the local governments within those counties would have received more than the 145 percent limit and fewer would receive less than the 20 percent floor.

[b] These jurisdictions have not been included in the "under 20 percent" category.

recipients, or about one-third of the total, have their revenue-sharing entitlements determined directly by a limit (Table 6).

In some cases, the redistributions caused by the limits result in more units being forced against a limit; in other cases redistributions remove jurisdictions from constraints. For example, in Missouri, Saint Louis receives more than 145 percent of the per capita state average. When the "excess" is redistributed among the other jurisdictions, it pushes a dozen other units including Kansas City up to the 145 percent limit, more than doubles the number of towns and municipalities affected by the 50 percent of taxes and intergovernmental receipts restriction, and removes almost fifty townships and municipalities from being affected by the 20 percent floor. Similarly, reduction in entitlements needed to bring all units in Indiana up to the 20 percent floor itself pushes others down to that floor.

As might be expected, by limiting eligibility to those units allotted at least $200 per entitlement period, small governments are affected almost exclusively. Eight out of ten of those affected have fewer than 500 residents, and seven out of ten of these jurisdictions were in the states of Kansas, South Dakota, Missouri, Texas, Nebraska, or Oklahoma. For the most part, these jurisdictions were located in rural areas.

The 145 percent limit affected a combination of jurisdictional types. The largest cities in thirteen states were hit in 1972 and, all told, some thirty-four cities with populations over 50,000 were affected by this limit. (These included Boston, Saint Louis, Norfolk, Philadelphia, Baltimore, Hartford, Wilmington, Louisville, Detroit, Newark, and Providence.) For the most part, these cities had very high tax efforts, and in some cases they were hit partly because they lacked underlying and overlying governments, such as townships, special districts, and counties. Although many of these cities were regarded as being in severe fiscal difficulty, the 145 percent limit reduced their grants and boosted those of their suburbs, thus impairing the redistributive impact of the revenue-sharing program. In some cases, this effect was substantial. For example, the allotments for Saint Louis and Philadelphia were cut by 75 and 54 percent, respectively, while those of their suburbs (as well as those for unconstrained localities in the rest of the state) were increased by roughly 18 percent in both cases. Another large group of jurisdictions affected by the 145 percent limit were resort communities such as Provincetown (Massachusetts), Rehoboth (Delaware), Sun Valley (Idaho), Vail (Colorado), and Mount Desert (Maine). In all of these cases, the measured tax effort of the community was extremely high. In large part this situation reflects the fact that resident income is a poor measure of the fiscal capacity of a community with large amounts of taxable property owned by nonresidents. A number of industrial enclaves such as Teterboro (New Jersey), Industry (California), and River Rouge (Michigan) were affected by the 145 percent limit for much the same reason. A significant number of low-density, rural county areas and their local governments were also affected by the 145 percent maximum. However, low per capita income more than high tax effort was usually responsible for their inclusion.

For the most part, the 20 percent floor acted to boost the amounts going to do-nothing jurisdictions—those with very small or nonexistent tax burdens. Included in this category were the 429 units that levy no nonschool taxes, many of the 13,000 townships in the plains and midwestern states, and the villages of New York, five county areas, and some other governmental relics that have been slowly dying in the graveyard of fiscal federalism. Many of these atrophying units have seen their governmental functions assumed by states, school districts, county governments, and overlying municipalities. In some cases they exist more for the perpetuation of their officeholders and for party organizational purposes than for any services they provide their residents. In other cases they have evolved into little more than special purpose districts which maintain a few miles of local roadways or carry out functions explicitly defined and financed by the state government. The 20 percent floor also boosted the amounts going to the enclaves of the super-rich; Hewlett Bay Park Village on Long Island with a per capita income in 1969 of $18,815 and its relatively impoverished neighbor, Great Neck Estates, with a per capita income of only $12,128, are both aided by the floor. So, too, are such better known stomping grounds of the wealthy as Scarsdale (New York), Beverly Hills (California), Palm Beach (Florida), and Grosse Pointe (Michigan).

The restriction that limits the maximum size of a jurisdiction's entitlement to 50 percent of its adjusted taxes and intergovernment revenues was felt primarily by do-little governments. In roughly half of the cases, this constraint became effective only because the amount going to the jurisdiction was boosted by the 20 percent floor. The county governments which are affected by this limit were hit in some cases because this limit transferred shared revenue from their towns and municipalities to them.

The Distribution

It is difficult to analyze the distribution that results from the constrained formulas below the level of the state areas. First, there is the problem of selecting the most meaningful way in which to express the size of the revenue-sharing grant. While per capita values are most commonly used, some alternatives and the justification for their use are:

1. Revenue sharing per $1,000 of personal income as an index of the redistributive impact of grants
2. Revenue sharing relative to the jurisdiction's expenditures as an index of the impact of the grant on the recipient government's operations
3. Revenue sharing relative to the taxes or the revenues raised by the recipient as an index of the increase in discretionary resources represented by the grant
4. Revenue-sharing grant net of some assumed contribution by the jurisdiction to support the program as an index of the program's net benefits (Table 7).[18]

The tremendous interstate and intrastate variations that exist in the functional responsibilities accorded even units of government with the same labels represent a second reason why the distribution of revenue-sharing money is difficult to analyze. These variations make comparing the

[18] The difficulty in attempting to calculate such net benefits is, of course, that it is impossible to know where the $30.2 billion would have been spent in the absence of revenue sharing. Would taxes have been lower? If so, which taxes and what are the incidences of these taxes? Would federal direct expenditures have been higher? If so, which type of expenditures? Would new programs be started or old ones expanded? Would other grants-in-aid have expanded more rapidly if there were no revenue-sharing program? If so, which ones? As Table 7 shows, the answer to the "net benefit" question is largely a reflection of the assumptions one must make.

TABLE 7. Value of Net Benefit to Alabama from the Revenue-Sharing Program for 1972 Under Various Assumptions Concerning the Source of Funds Used to Support Revenue Sharing

Assumption	Net benefit (in thousands of dollars)
Proportional increase in all federal revenues	+32,303.2
Proportional increase in federal personal income tax	+36,241.5
Proportional reduction in all federal outlays	+ 8,408.0
Proportional reduction in all federal grants-in-aid	− 9,389.7

Source: Robert D. Reischauer, "On Evaluating Revenue Sharing," October 1973, table 10, mimeograph.

amounts received by different local governments a meaningless exercise unless some account is taken of the revenue-sharing grants provided to overlying and underlying governments. Unfortunately, there are no easy procedures to allocate the funds provided to overlying and underlying governments for their constituent jurisdictions. Conceptually, the allocation should be based on the distribution of the benefits of the programs supported by the revenue-sharing entitlement. For example, if the grant to a county were used to reduce taxes, then the county government's entitlement should be distributed among the underlying municipalities according to the specific tax reductions each experiences. Similarly, if a county service were expanded or a new one instituted as a result of revenue sharing, the county government's entitlement should be allocated to each municipality and township according to the benefits their residents receive from the service. In practice, this would be an extremely difficult task because it would entail an analysis of the uses to which revenue-sharing funds were put, as well as a treatment of the distribution of the benefits of these programs. A somewhat simpler method would be for the overlying jurisdictions' entitlements to be divided among their constituent units of local government according to a combination[19] of *existing* contributions to the taxes of the county or state and *existing* benefits from county or state programs; that is, to use the average rather than the marginal tax and expenditure figures for the allocation process. In general, the highly urbanized sections of a county or state contribute a disproportionate share of the tax revenues. Unfortunately no similar generalization can be made concerning service benefits. In some states, county governments service the unincorporated or less densely populated parts of their jurisdictions and do very little, if anything, for residents of central cities and

---

[19] The weights used in such a combination would depend upon the extent to which the revenue sharing resulted in tax cuts and expanded services.

TABLE 8. Relative Size of Shared Revenue by County Areas[a]
(U.S. average = 100)

Size	Per capita	Relative to personal income	Relative to nonschool taxes
By 1970 population (000):			
1,000 plus	112	96	66
500–1,000	97	84	79
250–500	94	90	105
100–250	93	96	122
50–100	91	106	155
25–50	96	124	191
10–25	106	148	213
Less than 10	118	166	185
By population per square mile, 1970:			
3,000 plus	125	107	67
1,000–3,000	93	78	73
500–1,000	92	86	99
100–500	89	94	127
50–100	95	119	171
20–50	107	142	186
10–19	113	152	183
Less than 10	122	157	168
By percentage of nonwhite population, 1970:			
50 plus	141	180	112
30–50	124	151	130
15–30	111	107	84
5–15	93	86	88
Less than 5	93	98	124

Source: Nathan, Manvel, and Calkins, Monitoring Revenue Sharing.
[a] This does not include prorated amounts to the state governments.

large municipalities. In other states where county expenditures are largely in the areas of welfare and health, central city residents may be the chief beneficiaries of the county government's services.

The difficulties inherent in these complex procedures have led to the use of population to allocate amounts accorded to underlying and overlying jurisdictions. A very extensive analysis of the distribution of shared revenue using this methodology is detailed in Monitoring Revenue Sharing. The findings of this analysis are summarized in Tables 8 and 9. For county areas, per capita grants are greatest in those places with the largest and smallest populations, those with the most and least dense populations, and those with the greatest concentrations of minorities. Relative to income and to nonschool taxes, however, revenue-sharing grants are most significant for the smallest counties, the least densely populated areas, and those with the greatest concentrations of nonwhites. Looking at the nation's most populous cities, one finds that these areas generally receive far more than their surrounding suburbs do.

The Strings and Incentives

The incentives arising from the formulas and from the program's regulations could affect not only the fiscal behavior of state and local governments but also governmental structures. The sections which follow describe the major incentives and examine the evidence—or lack thereof—that governments are responding.

General Tax Effort Incentives

For the most part, recipient governments can do little to influence the factors that affect the distribution of shared revenue. A state, county, or even municipality, in the short run, cannot do much to increase the size of its resident population or to change the relative income of its citizenry. Furthermore, it is doubtful that at current levels of revenue sharing any jurisdiction would want to take actions designed either to attract population or to lower the average income of its residents just to obtain a larger entitlement.[20] It can, however, alter the level of its taxes, for the distribution formulas reward states which raise their taxes at a rate faster than the national average[21] and individual local governments which raise their taxes more rapidly than the statewide norm. This characteristic of the revenue-sharing program was intended both as a reward to high-tax jurisdictions

[20] Taking steps along these lines is of little value unless the Office of Revenue Sharing (ORS) acts to update the figures used in the distribution. At the local level it does not appear that relative income and population figures will be revised very often—possibly only once a decade. However, the census undercount of nonwhites has spurred a number of efforts to have the ORS increase the population figures used to calculate the entitlements of jurisdictions with large concentrations of blacks. See *City of Newark* v. *George P. Schultz and Graham W. Watt*, U.S. District Court for the District of Columbia, Civil Action No. 74–549, April 5, 1974; and Robert P. Strauss and Peter B. Harkins, "The 1970 Census Undercount and Revenue Sharing Effect on Allocations in New Jersey and Virginia" (Washington, D.C.: Joint Center for Political Studies, Howard University, 1974), mimeograph.

[21] Strictly speaking, it is the government whose tax effort rises relative to the average that is rewarded. It is clear, however, that the ORS will be able to update the tax collection figures ($LTAX$, $S + LTAX$) far more frequently than the aggregate income data ($AGINC$). In the following discussion the term "raising taxes" is used in a relative sense. If, on the average, jurisdictions were reducing taxes, this term would then hold for a government that was reducing its taxes at a less-than-average pace. Most of what follows about tax increases would apply in a negative way to tax cuts.

TABLE 9. 1972 Allocations of Shared Revenue to Local Governments Serving Selected Major Cities and Related Areas

| | City-area amount ($) per capita | | Amounts as percentage of state average | | | | | | | | |
| | | | Per capita | | | Relative to local government expenditure[d] | | | Relative to personal income | | |
City	Total[a]	City government only	City area	Balance of SMSA	Rest of state	City area	Balance of SMSA	Rest of state	City area	Balance of SMSA	Rest of state
New York	27.13	27.13	126	56	94	99	59	131	123	30	108
Chicago	21.23	18.57	129	63	107	122	56	121	133	52	126
Los Angeles	23.34	11.32	125	92	97	116	97	97	114	88	101
Philadelphia[b]	22.77	22.77	145	56[b]	100	95	59	113	147	87	101
Detroit	28.85	24.42	171	72	93	146	67	103	181	60	103
Houston	15.66	12.16	106	66	96	129	49	103	88	58	106
Baltimore	26.39	26.39	145	90	84	100	108	95	177	87	79
Dallas	16.14	13.82	109	61	96	102	48	105	83	51	107

Cleveland	24.82	19.41	185	81	95	141	68	101	210	63	98
Indianapolis	16.41	14.97	112	60	102	91	70	104	98	58	104
Milwaukee	29.22	17.58	146	69	96	129	59	103	139	55	104
San Francisco	24.90	24.90	133	79	102	101	75	104	114	70	104
San Diego	17.24	9.06	94	70	101	120	79	100	96	78	101
San Antonio	16.26	13.06	110	36	95	211	190	99	127	33	99
Boston[c]	28.05	28.05	145	90	100	102	92	109	160	40	106

Source: Data in columns 1 through 5 is from Nathan, Manvel, and Calkins, *Monitoring Revenue Sharing*; data in columns 6 through 11 is from the U.S. Department of Commerce, Bureau of Census, *Governmental Finance 1971–72* and *1970 Census of the Population*, vol. 1, *Characteristics of the Population*.

a Except in the italicized amounts where the same amount appears for "city government only," the city-area total includes a population-based proration of shared revenue going to the overlying county government, or, in the case of Indianapolis, to the underlying townships.

b For this multistate Standard Metropolitan Statistical Area (SMSA), the amount shown here excludes the portions of the SMSAs outside Pennsylvania.

c Boston SMSA data refer to the four-county Massachusetts State Economic Area C.

d These numbers tend to overstate the city ratios and understate those of the suburbs because only the municipal, school district, and prorated county expenditures were attributed to the city area.

and as an incentive to keep governments from channeling excessive amounts of shared revenue into tax cuts.

The extent to which state and local governments will respond to this incentive is unclear, but the considerable variation that exists among state areas and among local governments in the size of the incentive should provide a good basis for cross-sectional analysis. A major determinant of this variation is the particular formula under which the state area receives its entitlement. Table 10 lists the amounts each state area would have received in 1972 under the Senate, House, and final versions of the act had its tax receipts been $1 higher. The tax incentives are larger for those states receiving their entitlement under the Senate formula. This inducement in the final act ranges from 12.4 cents for Mississippi to 1.3 cents for Ohio.

Of course, the benefits to a state resulting from an increase in taxes by a constituent jurisdiction are spread throughout the area—that is, they cannot be captured exclusively by the government which raises its taxes. For example, if a state government were to increase its taxes, two-thirds of the resulting increase in the revenue-sharing entitlement to the state area would be spread among its local governments. The state government would automatically capture one-third of any increase in the state area's entitlement attributable to an increase in the tax effort of a single municipality. Other local governments in the state could see their entitlements decreased or increased as a result of the municipality's action.[22]

For any individual local unit, the tax incentive can be sizable. Those 3,040 jurisdictions currently affected by the limitation that no unit receive a revenue-sharing grant exceeding 50 percent of its adjusted taxes plus intergovernmental revenues are assured of obtaining an added 50 cents in shared revenue for every dollar of increased taxes until the 50 percent limit is no longer effective or until the 145 percent ceiling is hit. Small communities whose tax efforts and personal income are low relative to

---

[22] The effect on the entitlement of local government j in state s if local government k in state s increases its taxes depends on the value of a rather complicated expression. If j and k are county areas, and state s receives its grant according to the Senate formula, then the amount received by county area j will increase when taxes in k are raised only if:

$$(X) \left[\frac{\frac{LTAX_{sj}}{(PCY_{sj})^2(PCY_{sk})^2}}{\sum_{\text{all } h} \frac{LTAX_{sh}}{(PCY_{sh})^2}} \right] < (Y) \left[\frac{P_s \frac{PCY_{US}}{(PCY_s)(AGINC_s)} \sum_{i \neq j} P_i \left(\frac{PCY_{US}}{PCY_i} \right) \left(\frac{S + LTAX}{AGINC} \right)}{\left[\sum_{\text{all } i} P_i \left(\frac{PCY_{us}}{PCY_i} \right) \left(\frac{S + LTAX_i}{AGINC_i} \right) \right]^2} \right]$$

where

X = the percentage of the total revenue-sharing grant allocated to state s

Y = the percentage of the total revenue-sharing distribution to local governments in state s allocated to county area j.

TABLE 10. Projected Effect of a $1.00 Yearly Increase in Taxes on Revenue-Sharing Entitlements for State Areas Under the Senate, House, and State and Local Fiscal Assistance Act Formulas, for 1972
(Amounts given in cents)

State	Senate	House	State and Local Fiscal Assistance Act[a]
Alabama	10.14	1.51	9.28
Alaska	4.95	1.63	4.53
Arizona	6.35	2.06	5.82
Arkansas	11.28	1.52	10.33
California	3.83	1.84	1.68
Colorado	5.67	1.87	1.71
Connecticut	3.56	1.71	1.57
Delaware	4.80	1.83	1.67
District of Columbia	3.30	1.67	1.53
Florida	5.90	1.62	5.40
Georgia	7.56	1.56	6.92
Hawaii	5.63	2.19	2.00
Idaho	7.94	1.97	7.27
Illinois	4.13	1.69	1.55
Indiana	5.73	1.65	5.25
Iowa	6.32	1.90	5.79
Kansas	6.03	1.69	5.52
Kentucky	8.98	1.62	8.23
Louisiana	9.35	1.93	8.56
Maine	8.18	1.99	7.49
Maryland	4.47	1.85	1.70
Massachusetts	4.46	1.92	1.76
Michigan	4.79	1.83	1.67
Minnesota	5.74	2.03	1.86
Mississippi	13.54	1.91	12.40
Missouri	6.11	1.52	1.39
Montana	7.46	1.98	6.83
Nebraska	6.46	1.82	5.91
Nevada	4.13	2.02	1.85
New Hampshire	6.32	1.68	5.79
New Jersey	3.72	1.66	1.52
New Mexico	8.89	1.97	8.14
New York	3.59	1.91	1.75
North Carolina	8.36	1.63	7.65
North Dakota	9.21	2.21	8.43
Ohio	5.15	1.40	1.28
Oklahoma	7.55	1.54	6.91
Oregon	5.78	1.79	1.64
Pennsylvania	5.35	1.70	1.55
Rhode Island	5.57	1.88	1.72
South Carolina	9.93	1.59	9.10
South Dakota	9.00	2.16	8.24

(*continued*)

TABLE 10. (Continued)

State	Senate	House	State and Local Fiscal Assistance Act[a]
Tennessee	8.78	1.54	8.05
Texas	6.55	1.50	6.00
Utah	7.80	1.94	7.15
Vermont	7.07	2.30	6.48
Virginia	6.17	1.61	1.47
Washington	4.99	1.89	4.57
West Virginia	9.60	1.73	8.79
Wisconsin	5.92	2.22	6.41
Wyoming	6.62	2.17	6.07

[a] This represents the figure provided under the appropriate formula scaled down to conform to the aggregate appropriation of $5.3 billion.

those of other jurisdictions in the county can have even greater marginal incentives if they are not constrained by a limit. Instances occur in which a dollar increase in tax collections will garner more than a dollar in added revenue sharing. It is not inconceivable that some ingenious communities who face such high-tax incentives and who have the necessary legal powers may be tempted to boost property tax rates and use the resulting revenue to finance a local property tax relief or welfare program. If benefits from the new program were distributed roughly in proportion to the added tax burden (assessed value), both the taxpayers and the taxing jurisdictions could have their incomes raised simply by channeling more resources through the public sector.

Of course, those jurisdictions whose allotments are determined by the 20 percent floor or the 145 percent ceiling have no incentive to increase taxes, nor have they any disincentive against decreasing taxes. One may, therefore, find that those areas receiving the minimum grant tend to use their revenue-sharing grants disproportionately for tax reduction. Until they are affected by the 50 percent of adjusted taxes plus intergovernmental receipts constraint, such action is costless to them.

Of course, a minority of recipient governments face extremely high or low tax incentives. A fairly normal county pattern is illustrated in Table 11, which provides the marginal incentives for cities and towns of Fairfield County, Connecticut. An added dollar of taxes would garner Monroe 19.6 cents more in shared revenue but boost wealthy Darien's grant by only 3.2 cents. Four jurisdictions face no tax incentive, while for one-third of the governments the value of this incentive is between 6 and 10 cents.

At the state area level, the limited size of the revenue-sharing program may act to weaken the response to the tax incentive. If the program were

TABLE 11. Increased Revenue-Sharing Allotments Resulting from a $1.00 Increase in Taxes for Fairfield County, Connecticut, in 1972[a]

Towns	Amount in cents
Bridgeport	0.0[b] (5.3)
Danbury	9.2
Newton	0.0[c] (6.5)
Norwalk	6.5
Shelton	10.2
Stamford	4.5
Bethel	0.0[b] (19.2)
Brookfield	14.2
Darien	3.2
Easton	7.5
Fairfield	8.4
Greenwich	3.6
Monroe	19.6
New Canaan	3.3
New Fairfield	16.8
Newtown	19.0
Redding	8.8
Ridgefield	9.9
Sherman	7.3
Stratford	0.0[b] (11.8)
Trumbull	12.6
Weston	4.5
Westport	4.6
Wilton	6.3

[a] These estimates do not take account of the increased amounts gained by the jurisdictions because the state and county areas' allotments would be increased. Figures in parentheses are the unconstrained amounts.

[b] At the 145 percent limit.

[c] At the 20 percent floor.

expanded tenfold—to $53 billion per year—the tax incentives would rise proportionately. In other words, Mississippi would have received an extra $1.24 for a $1.00 increase in the tax figure used to calculate the 1972 entitlements. Since the amount distributed during the five years of the program will rise from $5.3 billion in 1972 to an annual rate of $6.75 billion in the last half of 1976, the average incentive will increase by about one-quarter. Whether governments will alter their behavior at current or even at the future levels remains to be seen.

A review of state tax legislation passed in 1972–73 and that proposed for 1973–74 does not show any clear indications that states with the most to gain through the revenue-sharing program have been boosting their taxes at an abnormal rate.[23] 1974 will be the second consecutive year that state governments have, on balance, reduced their tax rates. While the data needed for a credible test of the tax incentive effect will not be available until late 1975, crude regression analysis does not reveal that the rate at which state tax collections increased during 1973 was strongly related to the benefits derived from the revenue-sharing program.[24]

Responses of local governments to the tax incentive may be affected by a lack of information, institutional constraints, and uncertainty. Some jurisdictions may not know how large this incentive is; others may be unaware that their entitlement is determined by a constraint. For the majority of recipients, however, there does seem to be a general realization that tax collections do have a positive impact on the size of a government's grant; in fact, some local officials have raised the specter of smaller revenue-sharing checks to fend off popular pressure for tax cuts. Whether this reflects a true understanding of the workings of the revenue-sharing formula or is merely a tactic to gain public support for the officials' preferred course of increased expenditures remains to be seen. Responses at the local level may also be muted in specific cases because the recipient government is constrained by state taxing limits or other institutional bounds. The political cost of tax hikes, even those compensated for by devices such as reduced charges, could also act to retard local responses to the tax incentive. This is especially true because the reward is of uncertain size (because it depends on the responses of other governments); furthermore, it can be reaped only when the ORS updates its data two years after the tax increase. The possibility that Congress may terminate or modify the program adds another element of uncertainty.

[23] See Federation of Tax Administrators, "Trends in State Tax Legislation 1972–73," Research Report no. 66 (March 1974); "Revenue Proposals in State Budgets 1974," Research Report no. 67 (March 1974); and Tax Foundation, "State Tax Prospects, 1974," (April 1974).

[24] The regression results were:

$$(\% TX) = -0.019 + 0.163(Elast) + 0.004(RSINC) + 0.004(\% PY)$$
$$(-0.35) \quad (2.32) \quad\quad (0.80) \quad\quad\quad (1.07)$$

$$R^2 = 0.12$$

where

$\% TX$ = percentage increase in state tax collections between the first quarters of 1973 and 1974

$Elast$ = fraction of 1973 state tax collections derived from the income tax as a measure of the elasticity of the state tax system

$RSINC$ = increase in revenue-sharing grant that would result from $1 higher taxes

$\% PY$ = percentage growth in total personal income 1972 to 1973.

Despite these considerations, there are scattered signs that some local governments are responding to the tax incentive aspect of the revenue-sharing program. For example, Maryland's local circuit-breaker programs which shielded the elderly from high property tax payments were administered through a simple reduction in the tax liability. A number of counties now are maintaining the high liability but providing the elderly with a welfare check to compensate for the high burden. This bookkeeping device acts to increase the tax collection figures and thus the revenue-sharing allotment.[25] Other examples can be found in the new concern local governments are showing for the determination of their taxes by the Bureau of the Census. Some squabbling has occurred regarding who should be credited with shared taxes. According to what has become known as the "Memphis rule," taxes collected by a county government for its municipalities are credited to the tax effort of the county unless the governor specifies otherwise. Such taxes are important in New York, Tennessee, Alabama, Louisiana, North Carolina, and Nevada. In Alabama, Governor Wallace made such a specification, and as a result Jefferson County's entitlement was reduced by $1.4 million while the city of Birmingham had $969,000 added to its grant.[26] There has also been a good deal of controversy about the methods the Bureau of the Census uses to adjust the tax figures of jurisdictions with dependent school systems for the taxes used to support education. In Virginia, sixteen counties filed a law suit contending that the educational tax adjustment is so constituted as to neglect certain of their taxes and, hence, it tends to understate their true tax effort.

The Income Tax Incentive

The House interstate distribution formula purposely provided states with an incentive to rely more heavily on state income tax receipts for their revenues (Table 12, column 1). At the margin, these incentives are available only to the nineteen states that receive their entitlements on the basis of the House formula. Yet even some of these nineteen states face no marginal incentives. For example, since Minnesota's state income tax collections already exceeded the limit of 40 percent of the federal income tax liability for the state, no increase in the state area's revenue-sharing entitlement for 1972 would have resulted from increased income taxes. If Wisconsin were to expand its state income tax collections even moderately, it would face a similar situation (see Table 4). On the other hand, Con-

[25] The State of Maryland was considering making the same type of modification in its new circuit-breaker program.

[26] William M. Kimmelman, *Revenue Sharing and Local Government: The Case of the Birmingham Planning District* (Birmingham: Bureau of Public Administration, The University of Alabama, 1974), pp. 21–22.

TABLE 12. Increased Revenue-Sharing Entitlements Resulting from the Raising of State Income Tax Collections by $1.00 for 1972 Entitlements (Amounts given in cents)

State	House formula $1.00 added income tax	State and Local Fiscal Assistance Act	
		$1.00 income tax substituted for other taxes	$1.00 added income tax
Alabama	6.10	0	9.28
Alaska	6.13	0	4.55
Arizona	6.11	0	5.82
Arkansas	6.12	0	10.33
California	5.39	4.93	6.61
Colorado	6.07	5.56	7.27
Connecticut	0	0	1.57
Delaware	6.11	5.60	7.27
District of Columbia	6.10	5.58	7.11
Florida	0	0	5.40
Georgia	6.04	0	6.92
Hawaii	6.09	5.58	7.58
Idaho	6.13	0	7.27
Illinois	5.80	5.31	6.86
Indiana	6.04	0	5.25
Iowa	6.06	0	5.79
Kansas	6.11	0	5.52
Kentucky	6.08	0	8.23
Louisiana	6.11	0	8.57
Maine	6.14	0	7.49
Maryland	5.94	5.44	7.14
Massachusetts	5.84	5.35	7.11
Michigan	5.80	5.32	6.99
Minnesota	0	0	1.86
Mississippi	6.13	0	12.40
Missouri	6.05	5.54	6.94
Montana	6.12	0	6.83
Nebraska	6.12	0	5.92
Nevada	0	0	1.85
New Hampshire	0	0	5.79
New Jersey	0	0	1.52
New Mexico	6.13	0	8.14
New York	5.00	4.59	6.34
North Carolina	6.00	0	7.65
North Dakota	6.14	0	8.43
Ohio	6.01	5.50	6.79
Oklahoma	6.12	0	6.91
Oregon	6.03	5.53	7.17
Pennsylvania	5.78	5.29	6.84

TABLE 12. (Continued)

State	House formula $1.00 added income tax	State and Local Fiscal Assistance Act	
		$1.00 income tax substituted for other taxes	$1.00 added income tax
Rhode Island	6.12	5.61	7.32
South Carolina	6.09	0	9.10
South Dakota	0	0	8.24
Tennessee	0	0	8.05
Texas	0	0	6.00
Utah	6.12	0	7.15
Vermont	6.13	0	6.48
Virginia	5.98	5.48	6.95
Washington	0	0	4.57
West Virginia	6.11	0	8.79
Wisconsin	5.89	0	5.42
Wyoming	0	0	6.07

ᵃ Scaled down to conform to $5.3 billion appropriation.

necticut, Nevada, and New Jersey would not gain unless they greatly expanded collections from state income taxes. This is because all have a considerable way to go before their state income tax receipts equal the guaranteed floor of 6.67 percent of federal income tax liability.

For the remaining fifteen state areas, there is an incentive both to raise income rather than other taxes when new state revenues are required and to substitute income tax receipts for other existing state-level taxes. The second column of Table 12 lists the amounts that each state would have received in 1972 if it had substituted $1.00 of state income taxes for $1.00 of some other state tax. The third column indicates how much each state would have gained if it had increased its state income tax collection by $1.00 while maintaining its other tax receipts. For the states receiving their allotment under the Senate formula, this incentive is reported in column 3 of Table 10; for the others, it is that incentive *plus* the income tax incentive.

Potentially the income tax incentive could be substantial. One reason is that it does not require any increase in overall tax burdens. A state can get something for nothing simply by changing its revenue structure. For example, had Ohio increased its income tax to the point where the revenues generated equaled 40 percent of its federal tax liability and then reduced its state sales tax by an equivalent amount, the state area would have garnered $74.4 million—or 35 percent more--from the revenue-sharing

TABLE 13. Maximum Potential Gain in 1972 Allotments Resulting from Substitution of Increased State Income Taxes for Other State Taxes[a]

State	Millions of dollars	Percentage of state area 1972 grant
California	75.0	13.3
Colorado	9.0	16.5
Delaware	1.6	9.9
District of Columbia	1.8	7.5
Hawaii	1.2	5.0
Illinois	69.9	25.5
Maryland	14.8	13.8
Massachusetts	17.6	10.6
Michigan	38.6	17.2
Missouri	26.1	26.6
New York	41.6	7.1
Ohio	74.4	34.8
Oregon	2.5	4.7
Pennsylvania	56.8	20.4
Rhode Island	4.5	18.7
Virginia	18.0	16.9

[a] Only those states which at the margin can benefit have been included.

program in 1972 (Table 13).[27] A second reason is that state governments are the only participants. The state can be sure that it will receive one-third of any increase in the revenue-sharing entitlement resulting from increased income taxes. The remainder will be spread among units of government which are supported partially by the state's grants-in-aid. To the state government, such increases may act as substitutes for an expansion in existing intergovernmental transfers.

To date there is no evidence that the income tax incentive has been a dominant or even a significant factor influencing state taxing behavior. Those with the most to gain from the revenue-sharing program by boosting their income taxes are behaving much like the other states—that is, political considerations appear to be in the forefront. Of the ten states that reduced their personal income taxes in 1973, four (Ohio, New York, Michigan, and California) were among those the revenue-sharing program would have rewarded had they maintained their income taxes and reduced their sales levies by an equivalent amount.[28] Similarly, of the twelve state

[27] Relative to the total taxes, or even to the annual increases in tax collection, these amounts are fairly small. For example, the $74.4 million that Ohio did not receive represented 3.4 percent of its tax receipts in 1972 and 15.3 percent of the increased tax collection between 1972 and 1973.

[28] See Federation of Tax Administrators, "Trends in State Tax Legislation," and Tax Foundation, "State Tax Prospects."

legislatures considering income tax reductions in 1974, New York, Pennsylvania, Hawaii, and Ohio will reduce the relative size of their revenue-sharing allotment if they act. On the other hand, three (Virginia, Oregon, and Delaware) of the four states which have adopted or are considering increases in their personal income taxes are among those listed in Table 13. Looking at the income tax collection data, one finds that on the average personal income tax collections rose faster during 1973 in those states which do not gain from this aspect of the revenue-sharing formula than in those that benefit.[29] Of course, such data hardly constitute a rigorous test of the impact of the income tax incentive, since they reflect state differences in income tax progressivity and economic growth as well as rate changes. They do, however, indicate that there has been no overwhelming response to the income tax incentive during the first years of the revenue-sharing program.

Fees and Charges Effects

Just as there is an incentive for recipient governments to increase their taxes, there is an incentive for them to reduce their reliance on fees and user charges. Neither the interstate nor the intrastate distribution formulas take into account nontax revenues of any sort. Yet, some state, and many local, governments rely heavily on such receipts to generate their needed resources (Table 14). In some cases, well over half of the revenues raised by local governments come from the fees and charges placed on such services as sewage, trash collection, hospitals, and education (Table 15). In communities where the profits or surpluses of municipally owned water and electric systems are used to finance other city services, a much larger fraction of "total receipts from own sources" than of "total revenues from own sources" is derived from nontax sources.[30]

In the future, when a state or community is faced with the choice of raising its user charges or financing the services from increased general tax receipts, the latter course may be chosen because an increase in taxes would be reflected in the area's revenue-sharing entitlement. This is particularly true for the charges whose incidence is little different from existing taxes. Water, sewage, and trash collection fees in homogeneous suburban communities may be roughly proportional to current property tax liabilities. In such areas, therefore, the revenue-sharing program may lead to the erosion, if not the elimination, of such charges.

[29] See Department of Commerce, Bureau of the Census, *Quarterly Summary of State and Local Tax Revenue, October–December, 1973*, GT 73, no. 4 (March 1974).

[30] An extensive discussion of the arguments for and against including fees, charges, and utility surpluses in the distribution formula's index of effort is available in Nathan, Manvel, and Calkins, *Monitoring Revenue Sharing*.

TABLE 14. Nontax Revenues from Own Sources as a Percentage of All Revenues from Own Sources of State and Local General Purpose Governments, for Fiscal Year 1971

State	State governments	Local general purpose governments	State	State governments	Local general purpose governments
Alabama	18.18	46.00	Montana	23.15	20.68
Alaska	58.43	47.73	Nebraska	21.54	30.98
Arizona	17.05	27.48	Nevada	11.63	45.72
Arkansas	13.68	66.00	New Hampshire	26.31	16.04
California	12.96	26.76	New Jersey	17.45	13.27
Colorado	23.86	32.90	New Mexico	26.14	45.82
Connecticut	15.29	9.98	New York	12.14	18.19
Delaware	21.65	36.11	North Carolina	14.90	16.78
District of Columbia	15.29	13.14	North Dakota	32.45	27.24
Florida	11.84	36.27	Ohio	20.42	40.01
Georgia	14.07	31.19	Oklahoma	26.36	44.10
Hawaii	20.66	18.00	Oregon	23.33	39.53
Idaho	15.04	39.11	Pennsylvania	10.82	21.56
Illinois	10.37	23.74	Rhode Island	15.50	8.71
Indiana	22.19	34.50	South Carolina	16.56	44.68
Iowa	18.63	36.05	South Dakota	31.59	20.90
Kansas	20.65	26.98	Tennessee	15.34	33.59
Kentucky	17.71	50.08	Texas	18.98	32.18
Louisiana	23.56	37.16	Utah	22.46	31.78
Maine	17.82	9.58	Vermont	20.64	15.59
Maryland	15.48	19.59	Virginia	19.64	18.86
Massachusetts	10.86	11.61	Washington	16.79	38.26
Michigan	15.75	37.35	West Virginia	14.00	46.95
Minnesota	19.12	38.27	Wisconsin	14.62	28.72
Mississippi	15.12	56.40	Wyoming	28.36	55.49
Missouri	13.89	32.16			

Sources: U.S. Department of Commerce, the Bureau of the Census, *Governmental Finance in 1970-71*; and unpublished data from the Bureau of the Census.

Another possibility is that local, general purpose governments which operate school systems will attempt to earmark such receipts for education. As was indicated earlier, at the intrastate level, school taxes levied by general purpose governments are excluded from consideration in the distribution formula. Therefore, if a local general purpose government could somehow earmark all of its fees and charges for the support of its schools and shift an equivalent amount of school property tax receipts into the general fund, it could, through a simple relabeling procedure, increase its revenue-sharing entitlement up to the level the entitlement would have reached had new taxes been substituted for all fees and charges.

Such responses to the revenue-sharing program may be few and far

TABLE 15. Nontax Revenues as a Percentage of General Revenues from Own Sources for Selected Cities of Over 50,000, for Fiscal Year 1971

City	%
Huntsville, Ala.	58.9
Tuscaloosa, Ala.	82.9
Mesa, Ariz.	49.5
North Little Rock, Ark.	77.6
Pine Bluff, Ark.	54.9
Fort Smith, Ark.	66.2
Long Beach, Calif.	59.3
Ontario, Calif.	50.0
Colorado Springs, Colo.	52.3
Orlando, Fla.	50.4
Tallahassee, Fla.	87.2
Columbus, Ga.	56.3
Champaign, Ill.	55.7
Owensboro, Ky.	83.7
Flint, Mich.	64.3
Pontiac, Mich.	62.2
Wyoming, Mich.	69.1
Bloomington, Mich.	60.3
Columbia, Mo.	50.0
Springfield, Mo.	53.8
Lincoln, Nebr.	51.3
Binghamton, N.Y.	57.9
Rome, N.Y.	60.5
Fayo, N. Dak.	56.6
Cincinnati, Ohio	65.6
Norman, Okla.	78.2
Eugene, Oreg.	57.5
Galveston, Tex.	67.6

Source: U.S. Department of Commerce, the Bureau of the Census, *City Government Finance in 1970–1971.*

between, however, for three reasons. First, state laws often specify the revenue source from which the local contribution to education must come. Even in areas where local governments have some freedom to choose their means of educational support, state school aid may be keyed to local educational tax efforts.[31] In such cases the reduction in state educational aid which would result from a substitution of revenue from charges for school taxes would probably far outweigh any increase in revenue-sharing entitle-

[31] In most states, a fair proportion of the school districts have their state aid determined by the maximum or minimum limits in the school aid formula. In such situations, the state aid formula would not counteract the incentive to support schools from fees and charges. Neither would this be true where expenditures, rather than taxes or tax effort, act as the crucial variable in the state aid formula.

ment. Second, even if a local government went through the scenario described above, it would gain only if the ORS recognized the shift. The current methods used by the ORS to estimate the amount of school taxes that must be subtracted from the total taxes of local jurisdictions with school systems may not automatically result in recognition of the substitution. Finally, in most of the nation, schooling is not provided by those general purpose governments receiving shared revenue; instead, independent school districts (which may not be coterminous with counties, municipalities, or townships) generally provide educational services. The opportunity for such responses, therefore, will be confined to the ten states where dependent school systems prevail.

If changes in behavior do occur with respect to fees and charges, the effect will appear only gradually. To date, the only evidence to indicate a diminished reliance on fees and charges comes from the South, where a number of jurisdictions have considered levying a tax on community-owned utilities and simultaneously reducing the utility charges by an equivalent amount. The ORS has indicated that this is permissible, and it will actually increase the measured tax effort of the community if the community has the statutory power to levy the tax, and if the tax is specified appropriately on consumers' bills.[32]

Effects on Governmental Structures

There are a number of incentives in the revenue-sharing program that may affect governmental organization. One of these has to do with special districts. In the 1960s the number of special districts (excluding school districts) has increased by almost one-third to 23,900.[33] In some states such special districts account for a substantial fraction of local government activity. But because only general purpose governments receive revenue-sharing grants and since only the taxes collected by such units are considered in the intrastate distribution formulas, there exists a bias against areas where many governmental functions are relegated to the special districts that levy taxes.[34]

[32] See Revenue Sharing Advisory Service, *Revenue Sharing Bulletin*, vol. II, no. 8 (March 1974), p. 8.

[33] 1972 Census of Governments, vol. 1, *Governmental Organization*, pp. 1, 4. These are the districts, authorities, boards, and other entities classified by the Census Bureau as independent local governments. There are, of course, many more dependent special districts.

[34] A total of 11,581 of the 23,885 special districts have property-taxing powers; the remainder finance their operations largely from charges and special benefit assessments. Ibid., pp. 10–12. If special districts levied an equal fraction of all taxes in each jurisdiction within a state there, of course, would be no bias no matter how important special districts were.

In response to the revenue-sharing program, local governments may therefore become less anxious to spin off functions to new or existing special districts. Some real or *pro forma* consolidation could also take place. Special districts with taxing powers may be reabsorbed by a coterminous general purpose local government so that the latter unit could count as its own those tax revenues now collected by the special districts.[35] The reunification in many cases could be designed more to fulfill the Census Bureau's definition of a dependent special district than to change the behaviors, powers, or responsibilities of the governments concerned. Alternatively, special districts which now rely on their own tax receipts could be transformed so that they depended upon intergovernmental grants from the general purpose governments which underlie or overlie them. Since a merger or a shift to interlocal transfers would increase the revenue-sharing entitlement of the general purpose government, a bribe representing some portion of this increase could be held out to the special district for its compliance.

Whether such changes could be accomplished without state approval or whether the incentives are strong enough to overcome the institutional rigidity of the current structures, however, is open to some doubt. In many cases, fairly powerful incentives, such as the attempt to evade local debt or taxing limits, have been behind the proliferation of special districts. But the incentives in the revenue-sharing formula have not gone unnoticed. In Maryland, for example, a number of special park and planning districts and water and sewer commissions, which in the past have levied property taxes, are being folded into county governments. While the impetus for this change predates the revenue-sharing program, one of the arguments used by advocates of consolidation was that the receipts from the taxes levied by the districts would be counted in the revenue-sharing formula if the special districts were abolished. Much the same argument was made in a committee of the Florida legislature which in 1974 considered abolishing that state's special hospital districts that levied an ad valorem tax.

A somewhat related incentive exists for those local governments affected by the limitation that no revenue-sharing grant may exceed 50 percent of the recipients' adjusted taxes and intergovernmental revenues. Jurisdictions of this sort face a strong incentive to absorb an independent school district if a coterminous one exists. This incentive arises because intergovernmental transfers provided for education can be counted in determining the limit even though school taxes cannot. Since school districts usually obtain

---

[35] A total of 6,048 special districts are coterminous with existing counties, cities, or towns. Ibid., p. 70.

a large fraction of their resources from federal and state grants, it is highly unlikely that the 50 percent limit would still affect any locality which successfully absorbed a previously independent school district.[36] Up to the point where the limit ceased to become effective, the new government would receive fifty cents for every dollar in school aid. In some cases, this could be a significant sum with which to bribe the school district for its acquiescence. Of course, in most cases only states have the legal power to change the status of a school district, and since either the state or county would stand to lose from its revenue-sharing check an amount equivalent to the sum gained by the expanded local government, the opposition of the losers, as well as legislative and bureaucratic inertia, may thwart such responses.

The revenue-sharing program also contains a strong incentive to retain dying general purpose governments and, if possible, to create new ones and revitalize the scores of existing nonoperating governments. This incentive arises out of the guarantee that all general purpose local governments receive at least 20 percent of the statewide average per capita local grant. As was mentioned previously, many local governments which have few responsibilities benefit from this floor. Among them are 423 townships and villages in New York, 1,038 in Ohio, 705 in Wisconsin, and 870 in Minnesota.

While from the standpoint of efficiency (or even humanity) it would probably be best if many of these governments were abolished and their functions assumed by large governmental units, the revenue-sharing program makes it in the interest of both the residents of such jurisdictions and of the other inhabitants of a county area to retain such do-little governments. This is because the money needed to boost the grants of such governments up to the 20 percent floor does not come from the county area's entitlement but rather from a statewide pool. Eliminating such governments in one county area and shifting their responsibilities to, say, the county government could only decrease the total amount of resources provided that county area by the revenue-sharing program. While the county area's governments would still see their entitlements reduced by the same amount to support grants to do-little governments elsewhere, they would not benefit by receiving some of this pool for themselves. (Strictly speaking, every county area's contribution would be reduced slightly because fewer governments would have to be brought up to the floor.)

In areas where there are no governments benefiting from the minimum guarantee, the optimal strategy is to create a layer of such governments.

---

[36] In fact, in the ten states where dependent school systems prevail, none of the governments operating school systems is affected by the 50 percent limit.

In effect, such a strategy would tax the other county areas in the state for the newly formed governments' benefit. While states probably would object to the creation of entirely new layers of government, it is not inconceivable that the unincorporated, less densely populated areas of many counties could quickly set up governmental structures that would receive state recognition.[37] If the county or the government now responsible for providing services to such an area agreed to provide interlocal grants-in-aid to the new unit so that this unit could support its services without imposing taxes, the new government would be entitled to the minimum revenue-sharing grant while the sponsoring government would not be hurt.

Neighborhood governments in large cities offer another conceivable area of response. Such governments presumably would be largely or entirely supported by intergovernmental transfers from the parent city. If such neighborhood governments could obtain the status of local general purpose governments they would be entitled to a minimum revenue-sharing grant. In this connection, it will be interesting to see the position that will be accorded the community councils to be established in Indianapolis' city–county consolidation, the Uni-Government experiment.

Since state legislatures are dominated by the representatives of the governments whose revenue-sharing entitlements would be reduced to support such minimum payments, it is questionable whether governmental proliferation will occur. In states such as Delaware, however, where a large city which is not coterminous with a county area has had its entitlement reduced by the 145 percent ceiling, such procedures could be considered as a way of allowing the central city—in that case, Wilmington—to get back what was taken from it. In states where jurisdictional patterns differ significantly across counties, there could be pressure to equalize the situation. For example, in Illinois where 369 townships are boosted by the 20 percent minimum guarantee, there are seventeen commission counties which never had townships. As a result, many Illinois counties now have no, or only a few, jurisdictions lifted by the 20 percent minimum, while other counties have as many as thirty-five benefiting from this floor. Since the areas where no communities receive the minimum are having their grant reduced by roughly 8 percent to support the grants given to the areas which have not pruned away their skeletal governments, a fairly good case can be made for some action to equalize the situation.

Before concluding this discussion on the possible effects of the revenue-sharing program on governmental proliferation, it should be pointed out

---

[37] Preliminary, unpublished figures from the Bureau of the Census on new incorporations indicate that, if such a response is occurring, it is not of overwhelming significance. In 1970, new incorporations numbered 88; in 1971, 82; in 1972, 82; and in 1973, 112.

that this program is by no means the only, or even the most significant, impetus of this kind. Most states have taxes which they share with their local governments. A desire to obtain a piece of these revenues has long been the motivating force behind many a new incorporation.

It is conceivable that the revenue-sharing program may exert a small influence on annexation attempts and city–county consolidation movements. It has been suggested that the revenue-sharing program may provide marginal local governments with the resources necessary to stay afloat, thus undermining some of the fiscal pressures which have forced them to coalesce into jurisdictions of a more efficient size.[38] In the other direction, some cities may begin to covet the unincorporated areas on their borders, since annexation of this population could, under certain circumstances, boost their revenue-sharing allotments.[39] The fact that the revenue-sharing entitlement of a newly combined or altered jurisdiction will be more or less than the sum of the constituent jurisdictions' previous entitlements is sure to be used as an argument by forces favoring or opposing a merger. However, considering the strength of the other factors that have influenced the outcome of annexation and consolidation efforts in recent years, it seems unlikely that the revenue-sharing program will have a major impact.[40]

In Massachusetts and the few other states where townships and municipalities have similar responsibilities, there may be attempts by some

[38] There is also some indication that the revenue-sharing program is acting to undermine the nascent regional "governments" generated by state actions and by the requirements of the Federal A-95 procedures and the restrictions imposed by various water pollution and planning grants. While fiscal stringency once forced counties and towns in some sense to work with—or at least through—these organizations, now they are better able to go their own way. The Government Accounting Office (GAO) study, "Revenue Sharing: Its Uses by and Impact on Local Governments," Office of the Comptroller General of the United States (April 25, 1974, processed), indicated that in its sample of 250 recipients, 68 reported that revenue sharing had "encouraged regional intergovernmental projects, programs, and cooperation," and three reported the opposite impact.

[39] Only those jurisdictions which both before and after annexation received their grant on the basis of the 145 percent or 20 percent limits could be assured of increasing their grants. The incentive is further weakened by the fact that §51.23 of the regulations provides that only certain minimally sized annexations and boundary changes will be recognized by the ORS. Relative to most annexations these thresholds are fairly high. See Department of Treasury, Office of Revenue Sharing, *Regulations Governing the Payments of Entitlements under Title I of the State and Local Fiscal Assistance Act of 1972* (revised Aug. 1, 1973).

[40] Some early indications of this will be provided from the Bureau of the Census' Annual Boundary and Annexation Survey. See U.S. Department of Commerce, Bureau of the Census, *1972 Boundary and Annexation Survey.* The GAO survey on revenue sharing reported six cases in which annexations, incorporations, or consolidations aided the revenue-sharing program (1973), p. 26.

governments to get their classification changed because of the revenue-sharing program. The division of the intracounty money into two separate pots, one for municipalities and the other for townships, means that there could be significant gains to some jurisdictions from such a shift in its classification. When compared to the other municipalities in the county area, a certain city may appear to have a low tax effort and a relatively wealthy population. When contrasted to the townships of the area, however, the picture may be different. An example of this is the city of Revere in Suffolk County, Massachusetts. If it had been considered a town, its revenue-sharing entitlement in 1972 would have been 13 percent, or $105,000, higher than it was. Of course, all such shifts would require state approval and often could entail changes in governmental form and some alteration of responsibilities. For these reasons, then, there may be very few, if any, successful attempts by localities to change their governmental status.

Grants-in-Aid Effects

The revenue-sharing program may produce some peculiar responses with respect to grants between state and local governments and even to grants among localities. Among the former is the possibility that revenue sharing will cause state aid to local general purpose governments to grow far less rapidly than would have been the case in the absence of this program because states will regard revenue sharing as an adequate substitute for their own local aid programs. This attitude may be especially strong in the states where local revenue-sharing entitlements are large relative to the fiscal responsibilities of local governments. In such areas, the one-third of state area's entitlement reserved for the state government may represent a far smaller fraction of the state's current fiscal responsibilities than does the two-thirds reserved for the local governments (Table 16). Hawaii, North and South Carolina, and Delaware are some of the states where one might expect to find a marked dampening of the growth in state aid to local jurisdictions. Extreme responses of this type to the revenue-sharing program, however, will not occur because the act provides penalties for any state which reduces its aid to localities below the level received by such governments in fiscal year 1972. (Special adjustment can be made where the state government assumes a category of expenditures previously borne by localities or in which localities are given new taxing powers.)

In the states where a substantial number of county governments are affected by the limitation that no recipient's grant exceed 50 percent of its adjusted taxes plus intergovernmental transfers, there is yet another reason why state governments may be reluctant to increase their aid to localities. Every additional dollar of aid they provide to an affected general

TABLE 16. 1972 Revenue-Sharing Entitlements Relative to the General Revenues from Own Sources of State and Local Governments

State	State governments	Local governments	Index of neutrality[a]
Alabama	3.48	11.24	30.95
Alaska	0.90	5.07	17.74
Arizona	2.65	7.04	37.73
Arkansas	4.46	13.29	33.55
California	2.86	4.43	64.68
Colorado	2.69	5.50	48.97
Connecticut	2.39	4.65	51.36
Delaware	2.27	9.91	22.94
District of Columbia	0	4.66	0
Florida	2.72	5.78	47.00
Georgia	3.17	7.51	42.19
Hawaii	1.69	11.38	14.88
Idaho	3.22	9.18	35.13
Illinois	2.61	5.64	46.21
Indiana	2.80	5.54	50.57
Iowa	3.19	5.85	5.56
Kansas	3.00	5.56	53.90
Kentucky	3.90	11.01	35.37
Louisiana	3.21	13.12	24.49
Maine	3.72	9.99	37.20
Maryland	2.61	6.31	41.43
Massachusetts	3.27	5.72	57.28
Michigan	2.48	5.81	42.64
Minnesota	2.61	6.16	42.39
Mississippi	4.92	16.20	30.34
Missouri	3.30	6.69	57.96
Montana	3.87	6.55	59.01
Nebraska	3.45	5.24	65.86
Nevada	1.97	3.79	51.92
New Hampshire	3.44	5.75	59.76
New Jersey	3.06	4.47	68.41
New Mexico	2.67	14.24	18.73
New York	2.76	4.92	56.10
North Carolina	2.97	13.76	21.62
North Dakota	3.51	9.21	38.15
Ohio	3.20	4.90	65.36
Oklahoma	2.67	8.60	31.29
Oregon	3.05	5.82	52.37
Pennsylvania	2.67	6.68	39.96
Rhode Island	2.50	8.15	30.70
South Carolina	3.44	16.02	21.45
South Dakota	4.52	7.65	59.04
Tennessee	3.79	9.00	42.16
Texas	3.05	6.65	46.89

TABLE 16. (Continued)

State	State governments	Local governments	Index of neutrality[a]
Utah	2.94	9.77	30.09
Vermont	2.76	10.33	26.72
Virginia	2.74	7.79	35.14
Washington	1.92	5.86	32.75
West Virginia	4.63	12.71	36.47
Wisconsin	2.67	7.07	37.72
Wyoming	2.55	5.93	43.06

[a] Ratio of column 1 to column 2. The number 100 would indicate that the revenue-sharing program provides grants to the state and to all the local governments in proportion to their current fiscal responsibility. The lower the numbers are under 100, the greater the bias is toward localities.

purpose local government in such a county will not only increase the recipient government's revenue-sharing entitlement by fifty cents, but it will reduce the amount the state government receives by an equivalent amount. West Virginia, Kentucky, and Delaware are the states in which this type of response could be expected (see Table 5). Any slowdown in the growth of state aid to local governments resulting from the revenue-sharing program's bias toward local governments will manifest itself only gradually. No evidence is available yet to indicate that the states are considering the revenue-sharing windfall in their decisions regarding the expansion of state aid.[41] More likely than not, governors and state legislators will not act in a conscious way to counteract the effect of increased federal aid to local governments. Rather, they will respond to the decreased pressure from local grounds for larger state aid packages and to a general feeling that fiscal pressures on counties, municipalities, and townships have eased.

The existence of the 50 percent limitation could lead to the development of rather bizarre grants-in-aid between localities. For example, two

[41] The simple correlations between the growth in state aid from fiscal years 1972–73 and the ratio of revenue sharing to adjusted taxes and to adjusted taxes plus intergovernmental receipts are

Recipient	Taxes	Taxes plus inter-governmental receipts
All local governments	n.a.[a]	−0.12[b]
Counties	0.10	0.10
Municipalities	0.11	0.24
Townships	0.22	0.26

[a] Abbreviation n.a. means "not available."
[b] Revenue sharing relative to general revenue.

contiguous local governments, one of which is affected by the limitation, may agree to contract with each other for services. The unaffected locality could provide an intergovernmental grant to the affected jurisdiction in return for the latter's collecting the trash in both communities. Since the affected jurisdiction would receive an added fifty cents in revenue-sharing money for every $1.00 of increased intergovernmental transfers, it could provide the services at a bargain rate and still come out ahead.

A possible response to the situation in which both local and county governments found themselves hit by the 50 percent limit would be for the county government to increase or initiate aid to the townships and munici-palities within its borders. By so doing it can increase the amount of revenue-sharing funds provided to its constituent governments by fifty cents for each $1.00 of aid without reducing the level of its own revenue-sharing grant. In such instances, one could expect that governmental responsibilities once performed by the county government would be spun off to townships and municipalities.

In instances in which the towns and municipalities are affected by the 50 percent limit but the county government is not, just the opposite response could take place. Counties would have an incentive to reduce their aid, if any, to their constituent localities, for by doing so they gain fifty cents in revenue sharing for every $1.00 reduction in this aid. (When the limit affects such local governments, the excess is provided to the county government.) While the marginal incentives for this type of re-sponse are very strong, examples may be few and far between; because intergovernmental transfers from counties to towns or municipalities are rare, the opportunity will simply not be present.

While the preceding scenarios may seem a bit farfetched, they should not be dismissed. In 1972, 210 of the 279 units of local government receiving shared revenue in West Virginia were affected directly by the 50 percent limit; in Kentucky, the comparable figures were 260 out of 486. It is incon-ceivable that these jurisdictions are not considering means of reducing the impact of the constraints. Furthermore, old patterns and relationships have been changed by the revenue-sharing program. In at least one state, Maine, counties which never provided resources for their towns have undertaken new grants.

Effects on the Distribution of Public Service Responsibilities

As the preceding discussion pointed out, the revenue-sharing program is not a distribution of resources which leaves the relative strengths of various types of jurisdictions unchanged. Table 16 provides a measure of the bias between the state and all its local governments; it shows that, in general, local governments of the South are more strongly affected than those elsewhere. Table 17 provides a more refined breakdown for recipient types of local governments. In many instances, revenue sharing boosts the resources of townships and municipalities significantly more or less than those of county governments. These patterns are pronounced in North Carolina, South Carolina, Mississippi, Kentucky, and Pennsylvania. Even greater discrepancies occur among individual townships, counties, and municipalities, where revenue sharing can represent a 50 percent increase in a jurisdiction's total resources or an even larger expansion in its discretionary resources.

Such increases are of some significance because local governments are not free to use their grants in any way they desire. While state governments can spend their revenue-sharing money in any program area they wish, local governments are restricted to a list of "priority expenditures" which includes "ordinary and necessary maintenance and operating expenses" for public safety, environmental protection, public transportation, health, recreation, libraries, social service for the poor or aged, and financial administration, and "ordinary and necessary capital expenditures" in any program area (including education).[42] There are five other types of restrictions that must be observed as well. First, to ensure some visibility, the monies must be segregated in a separate trust fund and the nominal expenditures from these funds must be published periodically in local papers. Second, normal accounting and audit procedures must be adhered to. Third, activities or programs funded either wholly or in part by revenue sharing must be nondiscriminatory. Fourth, construction projects, more than one-quarter of whose support is derived from revenue-sharing monies, are subject to Davis–Bacon Act requirements, and those government employees whose wages are paid from revenue-sharing monies must receive compensation comparable to other government workers. Finally, recipients are prohibited from using revenue-sharing funds, directly or indirectly, as matching money for other federal grants.[43]

---

[42] The public safety category includes law enforcement, fire protection, and building code enforcement; environmental protection is a euphemism covering sewage disposal, sanitation, and pollution abatement; and public transportation covers expenditures on streets and roads as well as on transit systems.

[43] There is some confusion over just exactly what constitutes the "direct or indirect" use of the revenue-sharing funds to fulfill federal matching requirements. The regulations (*Federal Register*, vol. 38, no. 68, pt. 2, pp. 9137–9138) contain a very broad definition of indirect matching. However, the act and the regulations state that as long as a recipient's "revenues from own sources" for an entitlement period exceed the amount of such revenues collected in fiscal year 1972 there can be no violations of this provision unless the resources needed to meet increased federal matching requirements exceed the amount by which "revenue from own sources" rose. For a good discussion of this issue, see Otto Stolz, "Revenue Sharing," pp. 71–78.

TABLE 17. Shared Revenue as a Percentage of Adjusted Taxes (*T*) and Adjusted Taxes Plus Intergovernmental Revenue (*T* + *I*) for Recipient Local Governments, for 1972

State	Counties		Municipalities		Towns	
	T	*T* + *I*	*T*	*T* + *I*	*T*	*T* + *I*
Alabama	35.2	18.1	31.9	4.2	0	0
Alaska	17.9	1.7	13.6	5.9	0	0
Arizona	21.2	10.7	19.5	12.2	0	0
Arkansas	76.1	40.8	63.2	28.1	0	0
California	11.8	6.6	10.6	7.1	0	0
Colorado	21.7	8.5	16.5	11.0	0	0
Connecticut	0	0	11.5	6.2	13.3	6.3
Delaware	50.0	43.9	27.3	11.5	0	0
Florida	0	0	0	0	0	0
Georgia	16.3	12.1	15.4	14.3	0	0
Hawaii	22.9	17.4	21.9	17.5	0	0
Idaho	18.6	10.6	12.9	10.3	0	0
Illinois	35.8	21.0	33.0	22.8	0	0
Indiana	16.2	8.1	14.3	10.4	29.6	24.8
Iowa	21.2	10.1	17.9	11.8	42.4	40.9
Kansas	21.5	13.0	19.4	11.3	0	0
Kentucky	16.9	8.3	16.1	10.8	24.2	22.1
Louisiana	43.9	33.2	31.9	25.0	0	0
Maine	40.5	23.1	30.9	24.5	0	0
Maryland	24.8	23.0	21.5	12.9	22.6	14.3
Massachusetts	19.1	6.3	20.1	5.6	0	0
Michigan	11.5	10.0	11.4	7.4	11.0	6.9
Minnesota	19.8	7.3	16.1	10.0	29.7	15.5
Mississippi	20.5	7.5	17.5	9.9	32.7	18.4
Missouri	63.1	31.1	47.2	23.7	0	0
Montana	20.0	18.0	15.9	12.6	37.0	32.6
Nebraska	19.4	18.1	19.2	15.1	0	0
Nevada	21.0	7.8	17.9	12.5	29.2	28.0
New Hampshire	10.1	8.2	9.9	6.4	0	0
New Jersey	14.6	12.5	14.2	9.0	14.0	9.9
New Mexico	11.9	5.7	10.6	6.6	12.0	9.2
New York	55.1	27.7	49.3	16.7	0	0
North Carolina	10.1	5.0	9.3	7.0	12.0	9.8
North Dakota	37.1	5.3	28.2	19.5	0	0
Ohio	35.9	18.6	32.2	19.7	35.4	32.6
Oklahoma	16.2	8.2	14.9	10.3	33.4	20.1
Oregon	32.4	14.6	27.9	20.8	0	0
Pennsylvania	23.4	9.1	29.2	18.4	0	0
Rhode Island	22.4	17.1	15.9	10.6	21.1	16.3
South Carolina	0	0	20.2	10.4	20.5	8.0
South Dakota	66.6	33.8	47.4	36.3	0	0
Tennessee	24.5	22.3	22.3	19.8	28.9	22.8
Texas	30.9	8.2	22.9	8.6	0	0

TABLE 17. (Continued)

State	Counties T	Counties T + I	Municipalities T	Municipalities T + I	Towns T	Towns T + I
Utah	22.0	20.1	18.4	16.3	0	0
Vermont	29.4	24.1	30.1	26.0	0	0
Virginia	30.2	28.6	26.8	21.2	26.6	20.9
Washington	20.6	5.6	18.2	7.8	0	0
West Virginia	23.5	13.5	21.2	11.3	99.0	41.7
Wisconsin	49.8	44.5	41.9	30.0	0	0
Wyoming	22.1	8.0	20.6	7.9	38.2	8.6
	30.1	22.5	31.1	13.0	0	0

These restrictions should have only a minimal effect on large city and active county governments. This is because for such recipients, revenue sharing generally represents only a tiny fraction of the resources that are spent on nonfederally aided projects in the "priority expenditure" categories and constitutes only about half of the annual growth in outlays. For example, New York City's revenue-sharing entitlement for 1972 was smaller than the amount the city spent on health, fire, and environmental protection taken individually, and was less than one-half of the police department's budget; even the city's capital account for schools was larger. Between 1972–73, the city's tax collection increased by over twice the amount represented by its revenue-sharing allotment. This suggests that with only a bit of resourcefulness most large jurisdictions should be able to tuck their revenue-sharing grant into an existing expenditure program which meets the various restrictions and then to use the freed resources pretty much as they see fit—that is, for tax relief or for nonpriority items.

For thousands of small recipient governments, however, the restrictions can represent serious constraints which could affect their behavior. Not only is the revenue-sharing grant large relative to their current operations, but also many of these governments' powers are very circumscribed by tradition or state law. For example, many midwestern townships are little more than road districts; others provide some general assistance payments as well. Counties in such diverse states as Massachusetts, South Dakota, and South Carolina run courts and hire a sheriff but do very little else. Often the activities such governments are engaged in are either in no need of expansion, incapable of being expanded significantly, or not eligible as a priority expenditure (for example, general assistance). In the case of highways, the prohibition against use of revenue sharing for indirect matching of federal funds could pose a problem if federal aid is channeled into local roads.

Faced with this problem, many recipient governments will have three choices. First, they can spend their revenue-sharing funds on an expansion of unneeded or low-priority programs which they legally may provide and which the revenue-sharing program allows. During the initial years of the program this is likely to be their response, but after a certain period of time every local road in the nation will have been paved and resurfaced several times. The second option is to give the money to another unit of government—to provide it to a coterminous school district for capital outlays or to a county in return for the provision of certain specified "priority" services. Finally, the revenue-sharing resources could be used by the recipient to start new activities. Such action, of course, could represent a duplication of the efforts of some other level of government. But in the regions where some services such as fire protection and sanitation are not publicly provided, this may not be the case.

There is considerable evidence that problems of the sort described above have arisen. In Illinois, for example, townships are prohibited from engaging in many of the activities listed as priority expenditures. Those, such as roads, which do fall under both their purview and the "priority" label are incapable of absorbing the tremendous increase represented by revenue sharing. Faced with this dilemma the Illinois legislature enacted a law allowing "townships to contract with other governments, or individuals, or associations, or corporations for the . . . purpose defined as 'priority expenditures' under [the revenue-sharing law],"[44] provided that the services did not exceed their revenue-sharing entitlement. Since the townships were not empowered to use their non-revenue-sharing resources in this manner, the ORS issued an opinion that the new state law could have no effect because it violated section 123(A)(4) of the State and Local Fiscal Assistance Act which provides that revenue-sharing grants can be expended "only in accordance with the laws and procedures applicable to the expenditures of its own revenue." Similar situations have arisen in both Nebraska and Indiana. The probable upshot of this will be a series of state laws broadening the expenditure powers of certain classes of governments.

Miscellaneous Incentives

The detailed provisions of the revenue-sharing act may bring forth a number of other responses. A partial list of these would include changed accounting and budgeting procedures, increased citizen participation in expenditure decisions, a new concern by local governments over the Hatch

---

[44] Revenue Sharing Advisory Service, *Revenue Sharing Bulletin*, vol. II, no. 8 (May 1974), p. 3.

and Davis–Bacon acts and greater observance, or at least cognizance, of federal civil rights regulations. These topics, however, are beyond the scope of this chapter.

CONCLUSIONS

This review of the incentives inherent in the general revenue-sharing program make it clear that this is far from a neutral grant. While the "strings" and incentives are of a very different sort from those usually associated with categorical grants, the potential for altering, and distorting, the behavior of recipient governments is present in the revenue-sharing program.

Whether state and local governments' behavior will be significantly affected by the revenue-sharing program will become apparent only after several years. There are a number of reasons to expect responses to be muted. Institutional constraints may dominate. Taxing and debt limits, as well as state and local regulations specifying the functions, taxing powers, and expenditures permitted various types of localities, may limit the ability of recipient governments to respond to the incentives presented by the revenue-sharing program. The fact that the interests of different governments within a state can be in conflict with each other also may act to limit responses to some of the incentives. For example, in cases involving the 50 percent limit, state governments lose what their county governments gain; county governments lose what the townships and municipalities pick up. In situations involving the 20 percent floor or the 145 percent ceiling, the actions of one local government will increase or decrease the amounts received by other local governments. In cases involving tax increases all governments in a state area may gain, or some gain and others lose. The complicated network of impacts promises to politicize the responses of recipient governments to the incentives in the revenue-sharing program.

Finally, the uncertainty and lags involved may dampen the enthusiasm of recipients to "play" the revenue-sharing program. Actions taken by a jurisdiction today to increase its entitlement will be reflected only some two years later when the updated data are used in the allocation procedure. If the actions involve political costs—as raising taxes would—the promise of future benefits would have to loom very large to alter the recipient's behavior. Considering the possibility that the Congress may alter or even abolish the revenue-sharing program at the end of 1976 when the current act expires, few state or local governments are likely to embark on risky strategies to increase their entitlements.

4 Grants in a Metropolitan Economy—A Framework for Policy

ROBERT P. INMAN*

The fiscal problems of our metropolitan economies have in recent years become one of the major issues in domestic policy. Local political scandals, inefficient provision of services, and an inequitable distribution of both services and taxes are all facets of this fiscal crisis. Courts and legislatures at both the state and federal level have recently responded to these failings, the common remedy being grants-in-aid.

There are three primary forms of grants-in-aid—matching aid, fiscal base-equalizing grants, and revenue-sharing (lump sum) aid. *Open-ended matching grants* are transfers to local governments to cover a percentage of local expenditures on one or more public services. Federal aid for state and local welfare payments is the leading example. *Fiscal base-equalizing aid*, also known as percentage equalization aid, offers assistance to local governments to equalize taxing capacities across communities. Fiscally "rich" towns receive less aid than fiscally "poor" towns according to a well-defined, base-equalizing formula. The size of the base-equalizing grant is directly related to the level of local taxes.[1] *Revenue-sharing* or lump sum aid is a flat dollar transfer to local governments without explicit expenditure

*Assistant Professor of Economics and Public Policy, Department of Economics and School of Public and Urban Policy, University of Pennsylvania.

The financial assistance of the Spencer Foundation and the Fels Center of Government is gratefully acknowledged.

[1] In the fiscal base-equalizing scheme, the aid received by the local unit equals the difference between the revenue raised from its own fiscal base, at tax rate r, and the revenue it should have raised if its base were equal to the target base. If B equals the city's actual fiscal base per resident, B^* equals the target fiscal base per resident, and

and tax restrictions or incentives. Such aid may be related to community income, but its distinguishing characteristic is its intended "neutrality" with respect to local spending and taxation.

Much of the recent aid legislation involves combinations of these three primary forms of local assistance. The Federal Revenue-Sharing Act of 1972 is a case in point, building various open-ended incentive mechanisms for local tax reform onto a basically lump sum transfer program. (Reischauer and Pressman have examined this legislation in Chapters 2 and 3.) The purpose of this chapter is to extend the analysis to a more fundamental examination of the three building blocks of any aid policy. What are the likely effects of matching, base-equalizing, or lump sum aid on local expenditures and taxes? How might we evaluate these effects? And, finally, how can we use this information in a democratic policy process seeking to design a "best" grants-in-aid package?

A Policy Model of Grants-in-Aid in the Metropolitan Economy

Figure 1 outlines the central elements required of any policy model of the urban economy, detailed in this case for grants-in-aid. Three ingredients are needed. First, a list of feasible policy instruments must be specified. The list here includes matching aid, base-equalizing aid, and lump sum or revenue-sharing aid. Second, a behavioral model of the urban economy must be developed so that predictions can be made about the effects of various aid instruments on relevant policy outputs. Third, an evaluation mechanism must be defined so that policy-induced changes in the key policy outputs can be judged as an improvement or worsening of residents' well-being.

The subsection that follows describes one behavioral model of the urban fiscal economy's reaction to grants-in-aid. An important feature of this model, which distinguishes it from the previous work on the policy impacts of aid, is its general equilibrium specification.[2] Not only will aid have an

---

N is the number of residents, then Aid $= r(NB^* - NB)$. The city's tax rate under such a program is equal to the total tax requirement (T) divided by total fiscal base—$r = T/NB^*$. Substituting r into the aid formula gives, Aid $= (1 - B/B^*)T$. This is the usual percentage equalization formula, where $1 - B/B^*$ is a matching rate on local own revenues. For a generalization of this base-equalizing formula, see J. Coons, W. Clune, and S. Sugarman, *Private Wealth and Public Education* (Cambridge, Mass.: Harvard University Press, 1970), or R. Reischauer and R. W. Hartman, *Reforming School Finance* (Washington, D.C.: The Brookings Institution, 1973).

[2] For a review of the recent econometric work, see Robert Inman, "The Political Economy of Local Fiscal Behavior: An Interpretive Review." Paper presented for Ford Foundation Faculty Seminar on Urban Economics, held at Harvard University, 1974 (mimeographed).

GRANTS-IN-AID POLICY INSTRUMENTS	BEHAVIORAL MODEL	POLICY OUTPUTS	EVALUATION MECHANISM

Matching Aid (m) → Community 1 → $Y_{11} \cdots Y_{KF}$ (Private Incomes)

Base-Equalizing Aid (B) → Community K

Family 1

Social Choice Rule → Preferred Policy

Revenue-Sharing Aid (z) → Family F → $G_1 \cdots G_K$ (Services)

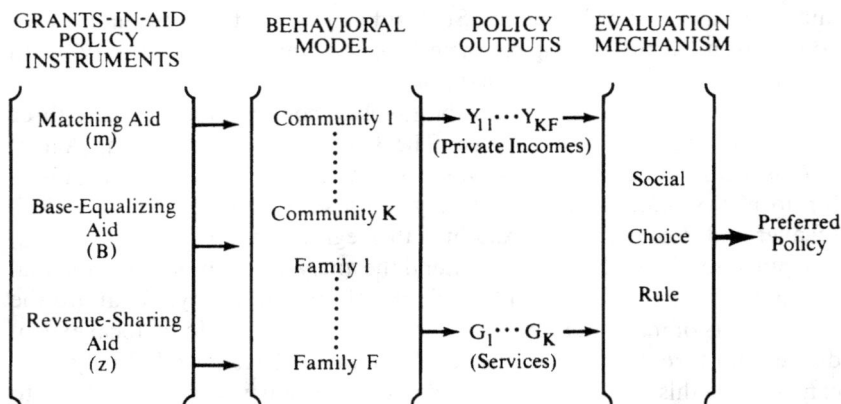

FIG. 1. A framework for fiscal aid policy.

initial, impact effect on local fiscal decision making, but important "second-round" changes may occur as well. The new post-aid configuration of public service outputs and local tax rates will induce a series of private market responses. People will move to newly favored towns and/or make adjustments in their housing stocks.[3] Such changes will affect the fiscal bases of communities. The behavioral model described below builds in such adjustments explicitly.

Finally, I discuss one general approach to the problem of evaluating the effects of aids policy. The specification is sufficiently broad so as to allow the policy model to apply three rather different evaluation rules: an equal public output (*Serrano–Hawkins*) rule; a sum of resident utilities (*utilitarian*) rule; and an "improve the worst-off" (*maximin*) rule.

A Behavioral Specification for Aids Policy

To analyze the impact of grants-in-aid in a regional economy, models of public service production, household location, housing and land value determinations, and local government fiscal behavior are required.

The output per resident of a local government (denoted by G) will be a function of the scale of public facilities provided by the local government (denoted by X) and the number of residents (N) sharing that facility. Formally, $G = (X, N)$. G provides consumption benefits to residents and will be an increasing function of X and a decreasing function of N (that is,

---

[3] C. M. Tiebout, "A Pure Theory of Local Expenditures," *Journal of Political Economy*, vol. 64 (October 1956).

congestion).[4] The facilities (X) which are used by residents are provided by capital and labor for which the local government pays a rental fee and wages. The total cost (C) of providing G units of output to N residents is denoted generally by $C(G, N)$. In the work which follows, I assume that (i)[5]: $C = k(G)N$, where the function $k(G)$ is the *cost per resident* of providing a given public service flow and is independent of the number of residents in the community. There is some recent econometric support for this assumption.[6]

To pay for the cost of local services, I will assume (ii) that local own revenues come only from the taxation of residential and commercial–industrial property. The effective local tax rate is given by:

(1)
$$r = \frac{C(1 - m) - Z}{BN}$$

where m is the percentage matching rate for matching aid, and Z is total exogenous or lump sum aid ($Z = z \cdot N$). The effective tax rate, r, is assumed (iii) to be proportional in all localities[7] but of course need not be equal from town to town.

In choosing a residential location, families are assumed to maximize their well-being, where satisfaction is a function of the private income (Y) and public services (G) which they can consume in each locality. As with the Tiebout model, I will assume (iv) that with the exception of residential land each community can offer the household the same bundle of private goods and service at the same prices.

The price of land in each community is determined by the competitive market for residential land and is specified as

(2)
$$p_L = p_{L_o}(PRIV) + \beta\Delta$$

---

[4] For a fuller treatment of this congestion specification, see Robert Inman, "The Specification and Estimation of a Shared Good Technology." Paper presented at the winter meetings of the Econometric Society, held in New York City (1973).

[5] The key assumptions of the behavioral model will be indexed by roman numerals.

[6] For a summary of the evidence, see W. Hirsch, "The Supply of Urban Public Services," in H. S. Perloff and L. Wingo, eds., *Issues in Urban Economics* (Baltimore, Md.: Johns Hopkins University Press, 1968); and T. Wales, "The Effect of School and District Size on Education Costs in British Columbia," *International Economic Review*, vol. 14 (October 1973).

[7] The theoretical analysis generalizes to a nonproportional local tax structure if the relative regressivity of the structure remains invariant to aid changes and the population movements induced by such changes. Thus, when reassessment occurs to adjust for the capitalization process in taxable property values, our analysis requires that the new assessments maintain the same relative tax structure through equal proportional adjustments in all assessed values.

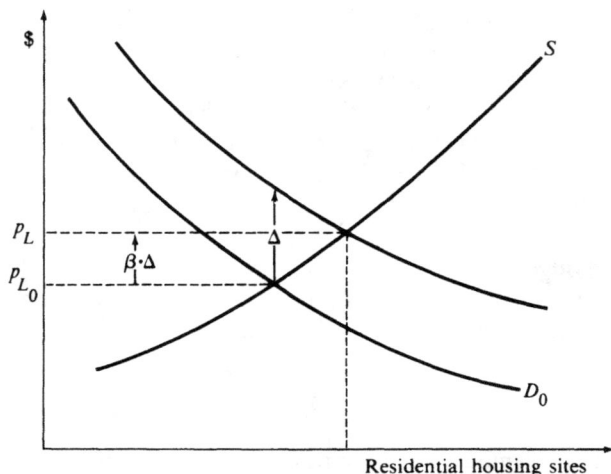

FIG. 2. The market for residential housing sites.

where p_{L_0} is a fictitious "base price" for land when all public services are centrally provided and depends on the private amenities (*PRIV*) of each community. To this base price we must add the market value of the fiscal advantages of each location, defined here as a capitalization factor, β, times the sum of the "willingness to pay," Δ (possibly negative) of all regional residents for entry into the community (Fig. 2). Generally Δ will depend on the level of public services and the tax rates in all communities while β will depend, in the usual way, on the slopes of the supply and demand schedules.[8] For this model, the community's fiscal attractiveness, Δ, is assumed (v) to be positively related to the level of its own per resident fiscal base, positively or not significantly related to the level of its own tax-financed G, negatively related to aid received by other localities and to the per resident fiscal base of other communities, and independent of changes in the level of tax-financed G in other communities.[9]

---

[8] In the case of linear supply and demand schedules, $\beta = \in_D / \in_{D_0} + \in_s$, where $\in_s$ equals the elasticity of the supply of housing sites and $\in_D$ equals the elasticity of demand for housing sites. For the derivation, see R. Musgrave, *The Theory of Public Finance* (New York: McGraw–Hill, 1959), pp. 291–292. The reader should note, however, that Musgrave's definition of β is for shifts in the supply schedule, while here β is specified for demand curve shifts.

[9] A more elaborate model allowing for interdependencies of public outputs would be a useful extension of this model. But if many communities are involved, the analytics of such a general equilibrium specification soon become intractible. An alternative compromise between reality and the demands of analysis might be a "two-community model" (center city versus suburb), applying oligopoly theory.

Given p_L, G, and r for their chosen community, households will select a stock of housing (H). All families within a town are assumed to reside on a single standard plot of land which costs p_L.[10] Of course p_L may vary across towns. Local governments tax both the value of land and housing. The gross price of housing is therefore defined by $p_h^* = (1 + r)p_h$, where p_h is the construction price of housing and is taken as exogenous for this analysis. A family's available income for spending on private goods and housing services (Y) equals its gross income (I) net of annual land costs and local taxes on the standard plot—that is, $Y = I - (\alpha + r)p_L$, where α is an amortization factor such that αp_L equals the annual cost of land. Gross family income (I) is assumed (vi) to be equal to wage–savings income defined by the market-determined rate of return (π) on the family's current (human and nonhuman) capital assets (A), plus the annual return on the family's land assets—$I = \pi(A + p_L)$. Residential land provides an investment as well as a place to live. Thus, $Y = \pi A - (r + \alpha - \pi)p_L$. The family's demand for housing services is specified by:

$$(3) \qquad\qquad H = H(p_h^*, Y),$$

where I shall assume (vii) that housing is a normal good ($\partial H/\partial Y > 0$) whose own price effect is negative ($\partial H/\partial p_h^* < 0$).

The tax base, b, of a typical family is the market value of its land and housing, $b = p_L + p_h H$. The taxable base of the community equals the sum of the values of commercial industrial property (M) plus residential property (Σb). Community fiscal base per resident (B) is therefore ($\Sigma b + M)/N$. In the work which follows we shall assume that (viii) there are a fixed number of housing sites in each community and that (ix) the market value of commercial industrial property is largely insensitive to local fiscal factors. Therefore, each town's N and M are taken as exogenous to our problem.

The level of local services and, correspondingly, the tax rate within each town are assumed (x) to be determined by a political process reflecting the desires of some decisive subset of community voters. The decisive resident or "resident-representative" has a downward-sloping, compensated demand schedule for public output defined by the continuous function,

---

[10] In effect, we are assuming a family's demand for land is a discontinuous function of price and income and that the aid policies considered here will not be sufficient to overcome these discontinuities and move the household to a new level of landownership. In a more cumbersome framework our model could be applied to a region with several zoning-defined land markets (an apartment market, a one-standard plot market, a two-standard plot market, etc.) in which fiscal reform would not be sufficient to move families from one market to another. While this latter characterization is closer to reality, no new conclusions would be gained over the simpler model presented in the text.

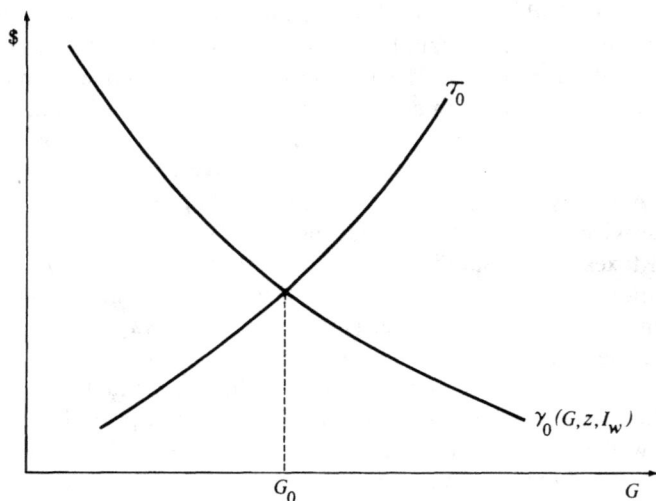

FIG. 3. Determination of local public output.

(4) $$G = g(\tau, z, I_w),$$

where z is the community's exogenous aid per resident ($z = Z/N$), I_w is the decisive resident's wage-savings income ($I_w = \pi A$), and τ is his marginal tax cost schedule. I shall assume (xi) G is a normal good ($\partial G/\partial I_w > 0$) and that increasing lump sum aid per resident will increase the decisive resident's demand for G ($\partial G/\partial z > 0$). The marginal tax cost schedule, τ, defines the extra tax dollars paid by a resident for an extra unit of public output and is a function of the level of local public services, the resident's share of the local tax costs (b/BN), and aid (m, z):

(5) $$\tau = \tau(G, b/BN, m, z),$$

where $\partial \tau/\partial G \geq 0$, $\partial \tau/\partial (b/BN) > 0$, $\partial \tau/\partial m < 0$, $\partial \tau/\partial z < 0$. The demand specification is completed by assuming $\partial G/\partial \tau < 0$ in equation 4.

Figure 3 defines the equilibrium level of public output, determined formally by substituting $\tau(\cdot)$ into $G = g(\tau, z, I)$ and solving for G. In Figure 3, the equilibrium is set by the intersection of the decisive resident's marginal benefit curve (γ_o, defined by the inverse of the compensated demand schedule (4)) and his marginal tax cost curve (τ_o).

Given these models of household decision making and local fiscal choice, how will grants-in-aid alter a community's provision of local services, as depicted in Fig. 3? A new equilibrium will emerge from a complicated network of household and fiscal interactions. Some sense can be made of the adjustment process if we impose one more restriction on our model of the urban political economy— an assumption of local political stability. For the remainder of this analysis I will assume (xii) that each district's decisive resident-representative before aid remains the decisive representative throughout the equilibrium adjustment process. When (xii) is valid, we can concentrate our disequilibrium analysis on this identified, decisive voter without having to predict a new decisive voter for each round of the adjustment process. Clearly this is a helpful simplification for our analysis, as the political decision-making process in each community is now "given." The assumption appears reasonable for most representative (nonreferendum) local choice processes for the policies we are considering here.[11] Relaxing this assumption would be an important extension of our work, and clearly we must consider the sensitivity of our policy conclusions to restriction (xii).[12]

Given assumptions (i) through (xii) we can now describe the regional economy's reaction to each of the three aid instruments. The flow diagram in Fig. 4 outlines the key linkages in the adjustment process for a particular community i. To conserve notation the decisive resident's variables (p_h^*, Y, I_w, b) and the community's variables (p_L, B, N, m, z, G, r) are both indexed by i.

Fiscal base-equalizing aid. Figures 4 and 5 describe the effects of base-equalizing aid. Figure 4 shows that the direct impact of changes in a community fiscal base per resident is to alter the decisive resident share of tax costs (b_i/B_iN_i). An increase in B_i, given N_i, will reduce the decisive resident's share of taxes, which shifts the decisive resident's marginal tax price schedule downward. This acts to increase G_i. These direct, first-round effects of base-equalizing aid are illustrated in Fig. 5 as a downward shift in τ from τ_o to τ_{DIR}. (To keep the diagrammatic analysis as simple as possible, I will assume that any income effects associated with changes in τ are small, thereby fixing the marginal benefit curve at γ_o. Our qualitative conclusions are in no way affected by this simplification. Income effects are allowed in our empirical analysis below). The point G_1 is the new preferred level of public services, but at G_1 marginal tax costs are increased over the

[11] Implicit here is the assumption that the resident–representative is insensitive to changes in the composition and tastes of those outside his winning coalition—that is, his coalition cannot be undone by changes in minority coalitions. In theory this assumption may be quite strong, but it accords well with the facts. See R. Lyke, "Representation and Urban School Boards," in H. Levin, ed., *Community Control of Schools* (Washington, D.C.: Brookings Institution, 1970); also, A. Vidich and J. Bensman, "The Clash of Class Interest," and R. C. Martin, "School Government," both in A. Rosenthal, ed., *Governing Education: A Reader* (Garden City, N.Y.: Doubleday, 1969).

[12] See, for example, Robert Inman, "Optimal Aid for School Fiscal Reform: A Simulation Analysis," University of Pennsylvania, 1975, mimeographed.

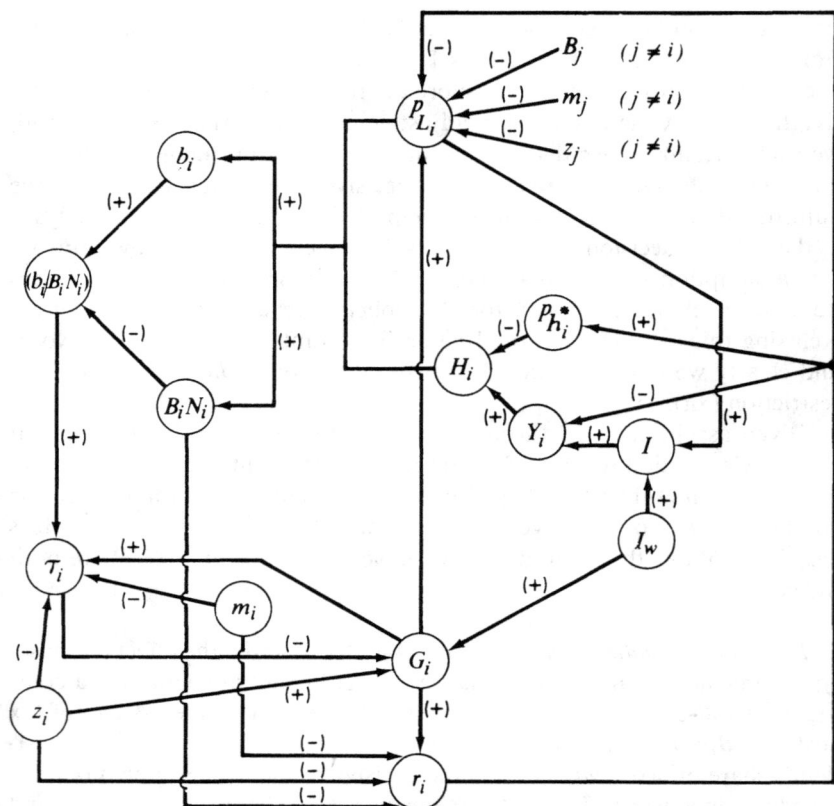

FIG. 4. The general equilibrium structure of local fiscal choice.

costs at G_o—the positive feedback from G to τ (from equation 5 and in Fig. 4)—and the adjustment process must move us to a lower level of G and τ.[13] The first-round stops at G_2, where marginal benefits now equal marginal tax costs for the decisive resident. At G_2 there is a new supporting tax rate, defined by equation 1, G_2, and the new fiscal base. As of yet, no families have moved nor has there been any adjustment in housing stocks in response to the initial changes in B_i, G_i, and r_i.

The second-round of the adjustment process begins when all families realize that B_i, G_i and r_i have changed in all localities receiving base-

[13] Stability of service provision in a regional economy requires this deflation effect. For a more detailed discussion of stability in this model, see Robert Inman, "Grants-in-Aid in a Metropolitan Economy: A General Equilibrium Analysis," University of Pennsylvania, 1975.

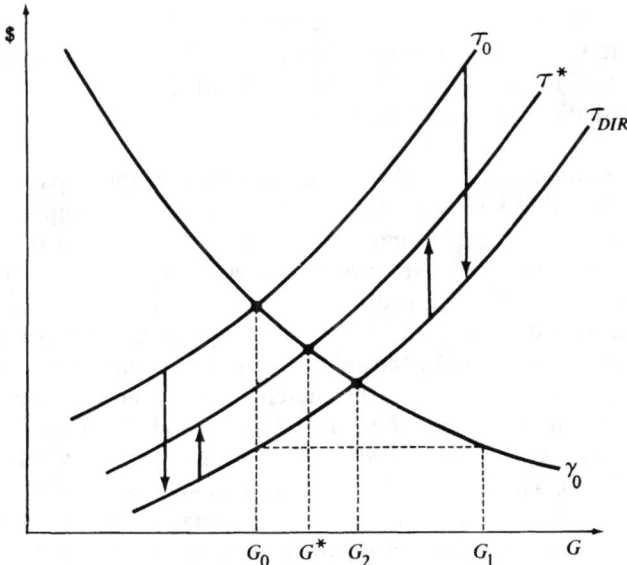

FIG. 5. The fiscal response to base equalization or matching aid.

equalizing aid. From the Tiebout adjustment process described by equation 2, we observe changes in the price of land, p_L, in each community. Towns which receive an increase in B_i (fiscally poor towns) will have increased G_i and/or reduced r_i, while towns which suffered a decrease in B_i (fiscally rich towns) will have reduced G_i and/or increased r_i. Aid-receiving towns ($dB_i > 0$) will therefore see an increase in their land prices while aid-contributing towns ($dB_i < 0$) will see a fall in their land prices. Changes in p_L will change Y for all residents through a wealth effect (πdp_L) and perhaps also through a renegotiated mortgage effect ($-\alpha dp_L$). Most likely the mortgage effect will be zero so that an increase (decrease) in p_L will increase (decrease) Y. The change in Y changes H, which, when coupled with the change in p_L, leads to a change in b for all families. Since base-equalizing aid fixes B_i at a target base for all communities, only b is allowed to change in this second round. The net effect is to change the decisive resident's share of tax costs (b_i/B_iN_i) in the same direction as the change in b_i.

The likely effects in the second round are a dampening of the first round changes. Communities which received base-equalizing aid will experience a rise in p_L, an increase in Y and H, and therefore a rise in b for the decisive resident. The rise in b_i will increase (b_i/B_iN_i) which leads to a reduction in G to G^*. This is shown in Fig. 5. The converse holds for communities which do not receive aid ($dB_i = 0$) or which contribute ($dB_i < 0$) to the base-equalizing scheme. The exact magnitude of the final change in G for each community will of course depend on a number of factors: the demand for public services, the actual degrees of fiscal capitalization which determine the changes in p_L, the rate of return on wealth (π), and the price and income elasticities of demand for housing.

Matching grants-in-aid. Like base-equalizing aid, the impact effect of matching aid will be to shift the decisive resident's marginal tax cost schedule downward for all communities receiving aid ($dm_i > 0$). The fall in τ_i leads to a first-round increase in G_i, shown in Fig. 5 as a rise in G from G_o to G_2. The second-round process is identical to the one described for base-equalizing aid, except now the community's fiscal base per resident (B_i) is not exogenously set by the aid program. For communities receiving differentially large increases in the matching rate (for example, poorer towns with redistributive matching aid), both b_i and B_i are likely to rise. The two changes work against each other in defining the new value of tax shares (b_i/B_iN_i). The final results depend, fundamentally, on the relative importance of commercial–industrial (M) property in B and the sensitivity of the market value of this property to changes in the fiscal configuration. Under the assumption (ix) that commercial–industrial property is insensitive to fiscal adjustments, those cities with large shares of their total base represented by M will likely see a rise in (b/BN) as b increases, or a fall in (b/BN) as b declines.[14] For these cities, the second-round effects will therefore dampen the first-round changes in G (Fig. 5). For cities with little or no commercial-industrial base, the changes in b_i and B_i will be nearly self-cancelling when defining tax shares (b_i/B_iN_i), and exactly so when $M = 0$ and all families alter b by an equal percentage. In these low M towns, the second-round changes in G will be small, with final equilibrium near G_2 (Fig. 5). Again the exact effects depend on the parameters of equations 1 through 5, where now the share of commercial–industrial property in the fiscal base plays a crucial role.

Lump sum (revenue-sharing) aid. An increase in revenue-sharing aid has two direct impacts on G_i. First and foremost is the direct impact that lump sum aid has on the budget constraint of the decisive resident. There is an outward shift in the constraint for an increase in aid ($dz_i > 0$) allowing the resident to buy more public and private goods. This "income" effect is

---

[14] In the simplest case, it is easy to show that if all families live on identical plots of land with identical houses, then the ratio b/BN can be rewritten as $b/(Fb + M)$, where F equals the number of families in the community. If M is fixed, then an increase in b will always increase this ratio while a decrease in b will reduce it. In the more general case of variable b's across families, the same conclusion holds as long as most family fiscal bases move in the same direction ($+$ or $-$) with policy changes.

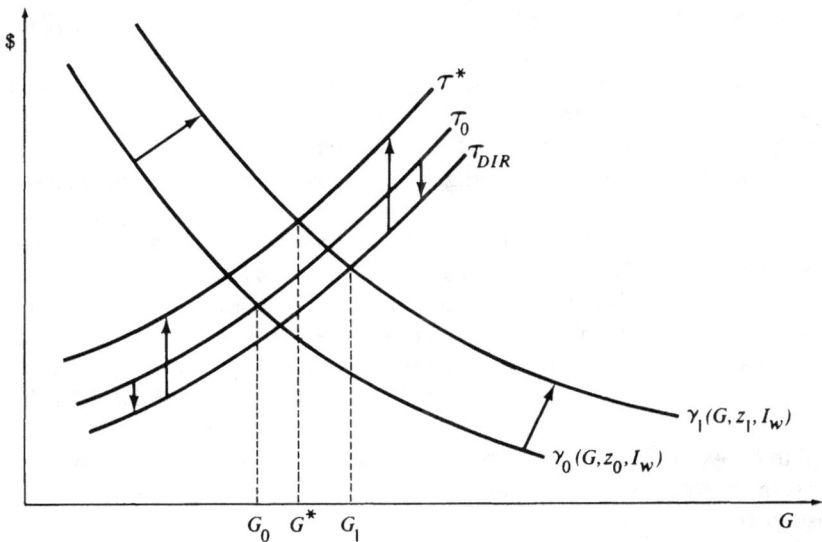

FIG. 6. The fiscal response to lump sum aid.

given by $\partial G_i/\partial z_i > 0$ from the compensated demand schedule (equation 4). Lump sum aid also alters the slope of the decisive resident's budget constraint leading to a small downward shift in the marginal tax cost schedule for an increase in aid (see equation 5). This effect also acts to increase G_i for an increase in aid, but it is likely to be of second-order importance. Figure 6 represents these two impact effects. The first-round outcome of an increase in lump-sum aid is shown as a rise in G from G_0 to G_1.

The analysis of second-round effects is identical to that for matching aid. Cities receiving differentially large amounts of revenue-sharing aid per person will see a rise in p_L followed by likely increases in b_i and B_i. As with matching aid, the net effect on the decisive resident's tax share (b/BN) will depend on the relative importance of the commercial–industrial (M) fiscal base. Where M is a relatively large share of the total base, there may be a significant dampening of first-round effects—for example, reducing G from G_1 to G^* (Fig. 6). Where M is unimportant, the final equilibrium values of G will be near the initial adjustment values—for example, G_1.

Financing aids policy. As each of the aid policies considered above involves the indirect transfer of taxable resources from one community to another, the initiating federal, state, or regional government must have a taxing mechanism to affect these transfers. For the purpose of this study, I will assume that the needed revenues are raised through a proportional tax on household wage and savings income (I_w) and that this tax has no appreciable incentive effects.

Introducing the effects of aid financing into our adjustment analysis requires us to trace through the equilibrium effects of changes in after-regional-tax earned income (dI_w) on the level of local services. Clearly, these adjustments to income changes are occurring at the same time as the system adjusts to the aid program, but it is convenient for expository reasons to treat each process separately. The initial effect of a regional tax on income is to shift the compensated demand curves downward for all decisive residents, implying a reduction in G_i in each community. As the tax is uniform across all communities in the region, there are no long-run incentives for families to relocate to other districts and therefore no adjustments in p_L. However, the decline in I_w will reduce Y, which leads in the long-run to a reduction in housing services, H. Thus there is a long-run fall in b. By our earlier reasoning, if the share of the commercial–industrial base in the total fiscal base is large, b/BN will fall, inducing a compensating rise in G.

For aid packages involving only lump sum aid, the equilibrium effects on G of our income tax-financed aid program can be computed by simply summing the equilibrium effects on an increase in z and the equilibrium effects of a decrease in I_w.

The equilibrium adjustment with regional financing is not so easily described for base-equalizing or matching aid however. For these policies, there is a potential simultaneity which must be resolved. For matching aid and base-equalizing assistance the level of total regional transfers will be a positive function of the levels of community spending on local services. As the G's increase, total aid will also rise. Therefore the required regional tax must change if the aid program's budget is to balance. Thus, for matching and base-equalizing aid the equilibrium pattern of I_w and G requires a simultaneous solution.

Figure 7 describes, heuristically, the solution to this problem. For each town there is an equilibrium Engel curve (EE) relating the equilibrium level of each town's G to various levels of after-regional-tax incomes, summarized by the town's decisive resident's level of I_w. In addition, for each town there is an equilibrium income curve (RR) relating the after-regional-tax income of the decisive resident to each town's level of G. Both curves are defined for each town, given B^* or the matching aid formula and the fiscal behavior of all other localities. A region-wide equilibrium occurs when all towns are providing a G ($= G^*$) defined by the intersection of their own equilibrium Engel and revenue curves.

There is no guarantee that a matching aid or base-equalization aid policy coupled with residential taxation will lead to a stable, long-run equilibrium

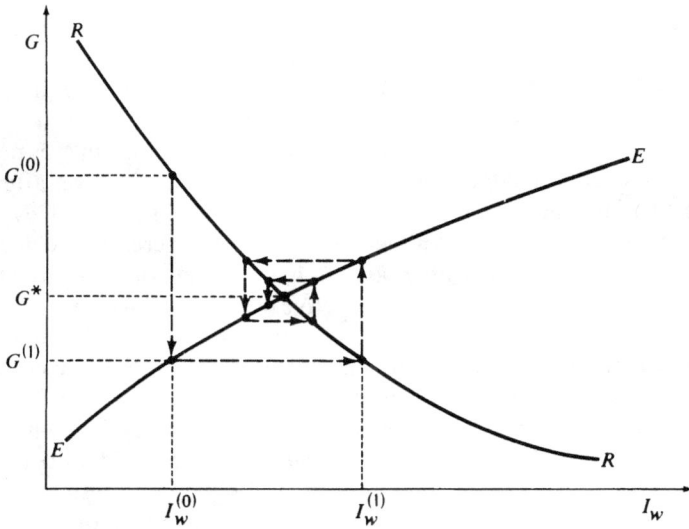

FIG. 7. The fiscal response to aid financing.

in the provision of local services. Stability clearly depends on the slopes of *EE* and *RR*; it is insured if the absolute value of the slope of *RR* is greater than the slope of *EE* for all towns.

A stable community is illustrated in Fig. 7. Given some initial pattern of public services, with $G^{(0)}$ corresponding to services in city *i*, there will be a corresponding break-even, grants-in-aid tax rate which will define an income level $I_w^{(0)}$ for the decisive resident in city *i*. The vector of new private incomes will lead to a new equilibrium vector of public output of which $G^{(1)}$ is a typical element. As the level of *G* has fallen in all cities, the aid program's tax rate will fall, leading to an increase in private incomes, say, to $I_w^{(1)}$ for the decisive resident in city *i*. In the next period, *G* is now increased in all cities, and I_w is subsequently reduced. The cycle spirals inward to a vector of G^*s if all district's *RR* and *EE* curves satisfy the stability conditions.

Theoretically, one unstable district will create instability in the whole system. But, in fact, a few unstable districts in a large region should leave the other districts largely unaffected, as their induced regional tax rate changes will be small.

An Evaluation Specification for Aids Policy

Having now specified the set of policy instruments (matching, lump sum, and base-equalizing aid) and how these aid instruments and their financing will influence local public services and residential after-tax income, we can turn to our final task in building a policy model of grants-in-aid: defining an evaluation rule for selecting among alternative aid packages. How shall we select a "best" grants-in-aids policy?

The approach adopted here assumes (xiii) that in selecting an aids policy, society should be ultimately concerned with the effect of aid on after-tax incomes (Y) and the level of local public services (G) received by each resident in each community. An increase in Y or an increase in G for one resident, holding the levels of Y and G fixed for all other residents, is assumed to increase the society's collective well-being. An aid program which gives such an increase in Y and G is preferred to one that does not. But more likely than not, aid programs will increase Y and reduce G for one person, reduce Y and G for another, and perhaps increase Y and G for a third. How do we judge these policies? Our evaluation rule must be able to weigh the tradeoffs between a single individual's Y and G and, at the same time, be capable of trading off Y and G between residents. One general evaluation rule which does perform such tradeoffs is the criterion: select those aid policies which maximize W, where W is defined generally as,

$$(6) \qquad W = \left\{ \sum_f U(Y_f, G_f)^\gamma \right\}^{1/\gamma}, \qquad \gamma \leq 1.$$

The function $U(Y_f, G_f)$ summarizes the planner-politician's evaluation of a given family's bundle of private income and public services, where f is a family index and the value of $U(\cdot)$ for each family are summed over all families. $U(\cdot)$ should be interpreted as incorporating both the family's own evaluation of its position as well as any social judgments on the household's level of Y and G.

To make the maximize W rule operational requires us to specify W exactly. $U_f(\cdot)$ must be given a parameterized functional form and a value for γ must be chosen as well. Recent debates over aids and tax policy have suggested three ways to formulate the objective function.

The Serrano-Hawkins objective. In a series of recent court cases, of which *Serrano* v. *Priest* and *Hawkins* v. *Shaw* are perhaps best known, the state and federal courts have been moving behind the wedge of the new "equal protection" doctrine toward a policy standard of equal local services.[15] Subject to the constraint that the metropolitan public economy remain a decentralized system of independent local governments (a constraint the courts hope will be nonbinding), the *Serrano–Hawkins* objective

---

[15] For a useful summary of these cases as they relate to the urban public economy, see "Equalization of Municipal Services: The Economics of *Serrano* and *Shaw*," *Yale Law Journal*, vol. 82 (1972), pp. 89–122.

is to make the distribution of G across residents as equal as possible. What happens to private income under the *Serrano–Hawkins* objective is not relevant. W therefore becomes a function of G only. To measure the inequities in the distribution of G, I shall adopt a variant of the recently proposed "Atkinson inequity measure"[16] denoted here by:

$$(7) \qquad\qquad W_S = \tilde{G}/\bar{G},$$

where $\bar{G}$ is the average level of G in the metropolitan region and $\tilde{G}$ is the distribution's, $f(G)$'s, "certainty equivalent" defined by $U(\tilde{G}) = \int U(G)f(G)dG$. The certainty equivalent $\tilde{G}$ of $f(G)$ is that level of G such that when each family receives $\tilde{G}$, society achieves the same level of welfare as obtained from the underlying distribution of G. When $U(G)$ shows the usual property of diminishing marginal utility, then W_S is bounded between 0 (complete inequality) and 1 (perfect equality).

The Utilitarian objective.   In contrast to the *Serrano–Hawkins* objective, the utilitarian form of 6 argues that both private income and public services are important when designing policy. W is therefore a function of both Y and G. In the utilitarian form, all individuals are treated equally, and social well-being is judged by the simple addition of individual well-beings, indexed by $U(\cdot)$. This is the classic Benthamite specification for judging community welfare. Equation 6 therefore reduces to:

$$(8) \qquad\qquad W_U = \sum_f U(Y_f, G_f).$$

Equation 8 follows from 6 with $\gamma = 1$.

The Maximin objective.   The maximin rule has been recently proposed as a policy alternative to the utilitarian criterion above. As derived by Rawls,[17] the maximin rule argues that in a just society resources should be allocated so as to make the welfare of the worst-off individual as large as possible. According to the Rawlsian theory of justice, "All social primary goods—liberty and opportunity, income and wealth, and the basis of self-respect—are to be distributed equally unless an unequal distribution of any or all of these goods is to the advantage of the least favored."[18] Operationally, we should identify the lowest level of family welfare as measured by $U(\cdot)$, and then select that policy where this minimum is maximized.

---

[16] A. B. Atkinson, "On Measurement of Equality," *Journal of Economic Theory*, vol. 2 (September 1970).

[17] J. Rawls, *A Theory of Justice* (Cambridge, Mass.: Harvard University Press, 1971).

[18] Ibid., p. 303.

To make the maximin rule operational, Atkinson[19] has shown that as the value of γ in equation 6 tends to minus infinity, policies which maximize W will be equivalent to maximizing the welfare of the worst-off family. Thus the maximin variation of our evaluation rule is to select those aids policies which maximize:

$$(9) \qquad W_R = \left\{ \sum_f U(Y_f, G_f)^\gamma \right\}^{1/\gamma} \quad \text{with} \quad \gamma \sim -\infty.$$

We have now outlined the main elements of a general equilibrium grants-in-aid policy model. The question of course remains as to whether this model or a variant of it can or should be used by decision makers. At minimum, we have to demonstrate that such models can be made to work for real-world policy problems. The following section summarizes one application of our methodology to a current policy issue; namely, designing grants-in-aid for support of local school spending.

An Application of the Aids Policy Model: The Reform of School Financing

Beginning with the watershed decision by the California Supreme Court in *Serrano* v. *Priest*, the present system of local school finance has come under increasing pressure for reform. Past complaints of inefficiency and racial discrimination have now been joined by new complaints of taxpayer inequities and fiscally induced inequities in the distribution of education across children. Grants-in-aid, in one form or another, have been the most prominently mentioned instruments for reform. Using the policy model described earlier, our objective will be to select an optimal grants-in-aid policy for a metropolitan region with many politically independent school districts.

To apply the aids policy model, each of its three components—instruments, the behavioral model, the evaluation rule—must be specified in detail. This specification is provided for a prototype metropolitan economy based on data from sixty-four school districts in Nassau and Suffolk counties, Long Island, and a "scaled-down" New York City.[20]

---

[19] A. B. Atkinson, " 'Maximin' and Optimal Income Taxation" (1972), mimeographed.

[20] The New York City student enrollment is reduced by one-third to keep the suburban–center city mix in reasonable proportions. Data for the simulations come from the *Fleischmann Report on Quality, Cost, and Financing of Elementary and Secondary Education in New York State*, vol. 1 (New York: Viking Press, 1972); the 1970 U.S. Census of Population; and the 1970 U.S. Census of Housing.

Specifying the Aids Policy Model

Four alternative fiscal reforms using grants-in-aid will be considered here. With the exception of family voucher proposals or fully centralized school spending, the plans outlined below cover the major reform proposals for school financing currently being discussed. Each builds on one of the grants-in-aid instruments analyzed in our policy model.

1. Lump Sum Aid with a Lower Limit (Foundation Aid, FA). This simplest of foundation aid proposals guarantees a minimum level of school spending per child. Each school district must spend at least this minimum, though it may buy more education (G) from its own fiscal resources if it desires. The relevant policy control variable is the level of lump sum aid per pupil (z).

2. Lump-Sum Aid with Upper and Lower Limits (FAUL). This proposal is identical to FA except each district's supplemental expenditure is now constrained to some percentage of the foundation mimimum. The FAUL proposal analyzed here sets the upper limit at a 10 percent increase over the foundation base. If the level of foundation aid is $600 per child, districts may spend up to $660 per child. The $60 comes from the district's own tax base. This proposal corresponds to the aid scheme recommended by the President's Commission on School Finance.[21] The policy variables are the level of lump sum aid (z) and the upper spending limit, set here at 10 percent.

3. Fiscal Base-Equalizing Aid (FBEA). The fiscal base-equalizing or "percentage equalization" aid proposal is a member of a general class of aid schemes known as district power equalizing aid (see item 1 above). FBEA has been recently recommended by Coons, Clune, and Sugarman,[22] and by some state courts as an alternative to the present system of school financing. The policy control variable for FBEA is the target fiscal base, B^*, which the grants policy insures.

4. District Matching Aid (DMA). The district matching aid proposal considered here is in the spirit of the present Federal Elementary and Secondary Education Act, Title I, where matching aid is distributed inversely to district income. For this analysis, the educational matching rate is defined as a linear function of mean district residential income, $m = \mu_0 - \mu_1$ (mean income). For positive values of μ_1, richer districts receive

[21] President's Commission on School Finance, *School, People, and Money: The Need for Educational Reform* (Washington, D.C.: Government Printing Office, 1972).

[22] John E. Coons, W. H. Clune, and S. D. Sugarman, *Private Wealth and Public Education* (Cambridge, Mass.: Harvard University Press, 1970).

a lower m. The regional matching rate is constrained to values between 0 and 1 for all districts. The relevant policy parameters for matching aid are μ_0 and μ_1.

To make the grants-in-aid behavioral model operational requires that we assign parameter values to the key production and behavioral equations, equations 1 through 5.

The tax rate equation (1) requires a specification of the district cost function for education. For this analysis, I shall assume all districts are equally efficient in the provision of education (G). G can therefore be measured in dollar units so that \$1.00 of school spending per child on instructional activities can be set equal to one unit of G. In addition, I assume the elasticity of educational costs with respect to the level of output (G) is ~ 1 (marginal cost equals average cost) for all districts.

The key parameters of the land value equation (2) are based on the econometric work in Oates' study of the capitalization of school spending and taxation in land values for northern New Jersey suburbs.[23] That study indicates an elasticity of suburban land values with respect to tax financed increases in education of about .10. As land values in the center city are likely to be less sensitive to educational changes because of restricted household mobility (both into and out of the city), an elasticity of .05 is assumed for New York City.

For the housing demand equation 3, we shall assume that for all families the price elasticity of demand for housing is 1 and that the marginal propensity to consume housing services from disposable income (Y) is $\sim .30$. These numbers are based on recent work by Aaron,[24] and De Leeuw[25] on the demand for housing.

For the demand-for-education schedule (equation 4), I assume the tax cost (τ), income (I), and lump sum (z) aid elasticities equal $-.12$, .24, and .25, respectively. The tax cost and income elasticities are based on an estimated Cobb–Douglas demand-for-education schedule for the school districts included in this study,[26] while the elasticity of G with respect to

---

[23] W. Oates, "The Effects of Property Taxes and Local Public Spending on Property Values," *Journal of Political Economy*, vol. 77 (November/December 1969).

[24] H. Aaron, *Shelters and Subsidies* (Washington, D.C.: Brookings Institution, 1972).

[25] F. De Leeuw, "The Demand for Housing: A Review of Cross-Section Evidence," *The Review of Economics and Statistics*, vol. 53 (February 1971).

[26] The tax cost variable was defined as $\tau = (1 - m)(b/B)\delta$, where δ measures the net dollar cost of one dollar of local taxes after deduction of local taxes from federal and state income taxes ($\delta < 1$ and decreasing in taxable income). The estimated demand equation for the Cobb–Douglas specification was:

$$G = 64\tau^{-.12}_{(.04)} I^{.24}_{(.06)} \qquad\qquad \bar{R}^2 = .38.$$

lump sum aid comes from the work of Ohls and Wales.[27] In defining this demand curve, I have assumed the decisive resident-representative corresponds to the resident in the 75th percentile position in each district's distribution of residential income.[28]

In defining the marginal tax cost schedule (5) for the decisive 75th percentile resident, I will assume that this resident deducts local school taxes from his state and federal income tax payments and that the share of total fiscal base in commercial-industrial property is .22 for our sample districts in Nassau County, .19 for the sample districts in Suffolk County, and .42 for New York City. These numbers are derived from the 1967 Census of Governments report.[29]

To complete the specification of our behavioral model of school district fiscal performance, we must allow for one more important choice—school districts may drop out of the public school system and provide education privately. Exit from the public system is likely to occur when participating districts are constrained by the aids policy to buy less or more G than their residents prefer or when residents face a locally-raised grants-in-aid tax which can only be avoided by dropping the local funding of education.

There are costs to exit, however. If a district drops public education, residents may no longer receive federal or state aid for educational expenses nor may they continue to deduct their school costs when calculating state

---

Standard errors are in parentheses. Both the price and income elasticities are significant at the .99 level for a two-tailed t test. A Chow test of the null hypothesis that New York City and the suburban districts behave according to the same relationship was accepted at the .99 level.

[27] J. Ohls and T. Wales, "Supply and Demand for State and Local Services," *The Review of Economics and Statistics*, vol. 54 (November 1972).

[28] The quantitative evidence on political participation and leadership in school decision making by income (or property) class is limited. What evidence there is does suggest that the decisive resident will generally be a member of the professional class, an established member of the community, and near the upper end of the district's wealth distribution. See, for example, W. Bloomberg and M. Sunshine, *Suburban Power Structures and Public Education* (Syracuse, N.Y.: Syracuse University Press, 1963), ch. VII. One technical point should be mentioned. When b equals some constant times I then the 75th percentile resident in the distribution of I will also be the 75th percentile resident in the distribution of b. The assumption above implies that this proportional relationship between b and I holds.

[29] See 1967 Census of Governments, *Taxable Property Values* (Washington, D.C.: Government Printing Office, 1968), p. 136. In the case of fixed residential sites (N fixed) and exogenous commercial fiscal base, the marginal tax price schedule, defined generally by equation 5, can be written explicitly as:

$$\tau = \delta(b/BN)(1 - m)C' + \delta r\theta \, db/dG,$$

where θ is the share of commercial property in total fiscal base and δ is the deductability effect ($\delta < 1$) defined in footnote 26 above.

or federal income taxes. In addition, even if a district exits, residents must still pay federal, state, or regional taxes (only local taxation is avoided) to support any aid to those districts remaining in the public system.

How does a school district balance these benefits and costs? For this specification, I shall assume that the decision to stay or leave is made by the decisive resident for his entire district on the basis of *his* own net benefits from the consumption of education through public or private schools,[30] that the drop-out decision is made after the first year the aids policy is introduced and is irreversible, and that the center city is legally or politically always constrained to provide a public school system. These assumptions yield an upper limit to the abandonment of the public system.

The final element in the specification of the grants-in-aid policy model is the definition of the evaluation rule. Each of the three objective functions given in equations 7 through 9 will be examined in our policy simulations. We can therefore test the sensitivity of our optimal aid conclusions to the underlying evaluation rule. Does one aid policy dominate all others over three, philosophically very different, policy objectives? To make rules 7 through 9 operational, I assume a Cobb–Douglas specification for $U(Y_f, G_f)$, where $U = A Y_f^{.83} G_f^{.17}$ for the utilitarian and maximin rules, and $U = BG_f^{.17}$ for *Serrano–Hawkins* rule. For W_R, γ is set equal to -10.[31] Other values for the exponents on Y and G should of course be tried, but for purposes of illustration these specifications of U seem reasonable.

Some Policy Simulations: The Search for an Optimal
School Aid Policy

The results of the policy simulations, using the model described above, are reported in Table 1.[32] The selection of the preferred aids policy proceeded in two steps. First, given an evaluation rule (W_S, W_U, W_R), the optimal value for each control variable (z, B^*, μ_0 and μ_1) for each of the

---

[30] Since all residents, whether they use public schools or not, must pay the regional income tax required to support fiscal reforms, our drop-out analysis can concentrate on the relative fiscal advantages of local tax versus private funding. The decisive resident will prefer that alternative which maximizes his family's net benefits from education. Net benefits are calculated as the area difference between the marginal benefit and marginal cost curves for each of the alternative public and private financing schemes. See Inman, "Optimal Aid," for details. The relevant net benefits are calculated using $E = 64\tau^{-.12} I^{.24} z^{.15}$ as the specification for the demand curve. Families in districts dropping public education are assumed to purchase education from the private system consistent with this demand schedule.

[31] $\gamma = -10$ was as large a negative number as could be accurately used by the computer when solving for W_R.

[32] The reader is referred to Inman, "Optimal Aid," for a detailed discussion of the results in this section.

four aid policies (FA, FAUL, FBEA, DMA) was selected.[33] These results appear in each column of Table 1. For example, from among all possible values of lump sum aid, z = $3,000 per pupil is the value of FA which maximizes W_S. Second, for each evaluation rule, I compared the results from each of these best-aid programs and selected that aid policy from among the four which achieves an overall maximum of the objective function. That aid program—the "best of the bests"—is the optimal policy for the given objective. For example, FAUL policies with z = $550–$860 per child is the optimal aid program for the *Serrano–Hawkins* equal output objective. These *maximum maximorum* policies are indicated by an asterisk.

The preferred policy by the *Serrano–Hawkins* criterion is foundation aid with an upper limit. Perfect equity (W_S = 1) is achieved with any foundation level above $550 per child and below $860. The $550 lower limit proves sufficient in our specification to hold all districts in the public school system, thereby insuring federal or state control over local school spending. (The upper level government is assumed to have no direct control over family spending on private schools). The $860 upper bound is the maximum at which aid can be set before some districts choose to spend less than the allowed 10 percent supplement. Between the $550 and $860 limits all districts are in the public system and are spending the maximum allowed under the supplemental spending constraint. Thus we achieve perfect equality of school spending. Foundation aid with only a lower limit may be a reasonable alternative, but for equality to be achieved (essentially by forcing all districts to this foundation level) very high foundation levels are required. DMA and FBEA policies are distant third choices. As is so often the case in public policy decisions, we see in these results a confrontation of equity and efficiency or, in the language of the courts, equity and local control. DMA and FBEA give maximum local choice on fiscal matters but are poor equity performers. The preferred equity policies minimize local control.

The optimal aid program under the utilitarian and maximin criteria is a foundation aid plan with the foundation level set at $690 per pupil and

---

[33] The search for optimal z values for FA and FAUL proceeded in $10 intervals from zero to $5,000. The search for optimal B^* levels for FBEA was made in $1,000 intervals from zero to $500.000 per pupil. Optimal value of μ_0 and μ_1 for DMA proceeded in a search of intervals of .05 between -3.0 and $+3.0$ until a maximum was found, and then in intervals of .01 around that maximum. Local maxima did not prove a serious problem in our search for optimal parameter values.

For DMA and FBEA the system converged rapidly to equilibrium values of G for regional income tax financing. The convergence criterion was for the average absolute deviation of $G^{(t)}$ from $G^{(t-1)}$ to be less than five dollars. Convergence was achieved in two or three iterations in all cases, and for most districts the final value of G was within one or two dollars of the previous iteration's value.

TABLE 1. School Financing and Optimal Aid Programs

	Serrano-Hawkins objective				Utilitarian objective				Maximin objective			
	FA	FAUL*	FBEA	DMA	FA*	FAUL	FBEA	DMA	FA*	FAUL	FBEA	DMA
Optimal policy parameter	$z =$ \$3,000	$z =$ \$550–860	$B^* =$ \$28,000	$\mu_0 = 1.55$ $\mu_1 = .06$	$z =$ \$690	$z =$ \$1,250	$B^* =$ \$88,000	$\mu_0 = .05$ $\mu_1 = 0$	$z =$ \$640	$z =$ \$1,250	$B^* =$ \$29,000	$\mu_0 = .17$ $\mu_1 = 0$
Regional tax rate	.38	.07–.11	−.013	.049	.088	.159	.077	.002	.082	.159	−.01	.007
Value of objective function	$W_S =$ .9995	$W_S = 1.0$	$W_S =$ .98	$W_S =$ .9954	$W_U =$ .9999240	$W_U =$ .9999230	$W_U =$ .9999221	$W_U =$ .9999222	$W_R =$ 758.75	$W_R =$ 742.58	$W_R =$ 722.02	$W_R =$ 738.74
Mean G (dollars)	3,008	605–946	596	569	1,028	1,354	949	626	995	1,354	601	611
Coefficient of variation	.01	0	.19	.09	.10	.03	.40	.14	.09	.03	.20	.14
Center city G Suburban G	~1	1	.76	.98	.97	1.03	.57	.81	~1	1.03	.76	.85
Percent enrollment in private schools	0	0	.08	0	0	0	0	.24	0	0	.08	.17

$640 per pupil, respectively. Foundation aid with the 10 percent supplemental spending limit is a close second. DMA and FBEA are not serious contenders under either the utilitarian or maximin objectives.

How can we explain this consistently superior performance of the lump sum, revenue-sharing aid proposal under the utilitarian and maximin criteria? First, aid policies such as DMA and FBEA, which operate primarily by shifting the tax cost schedule, τ, prove to be ineffective instruments for altering the distribution of educational resources for this region. The low price elasticity of demand $(-.12)$ coupled with the offsetting capitalization process described earlier severely limit the impact of these policies. Districts which receive relatively favorable price reductions with a rise in m or B see these gains reduced through the partial capitalization of the advantages into their decisive resident's fiscal base (b). FA and FAUL, which affect all districts equally, are not as limited by the capitalization process. In addition, FA and FAUL operate through changes in lump-sum aid, to which districts seem more sensitive.

Second FA dominates FAUL on allocative efficiency grounds as districts are free to choose their own preferred G levels above the minimum. But FA retains sufficient control through its minimum spending constraint to respond, along with FAUL, to the heavy equity demand of a maximin criterion. Of the two policies which work, FA is more flexible.

On balance, it appears that lump sum aid proposals dominate both base-equalizing and matching grants as policy instruments for improving the financing and provision of education in this prototype metropolitan economy. Whether these results will hold true for other urban political economies and other public services must remain an open question. There is tentative evidence to suggest that the results for education are fairly robust across alternative behavioral specifications of the urban economy.[34]

While the conclusions from Table 1 are suggestive, the main point of these simulations is not the exact numbers but the fact that a formal grants-in-aid policy model can be constructed and used to produce reasonable

[34] Testing the sensitivity of these results to alternative price and income elasticities of the demand for education $(-.4$ and $.65$, respectively) while holding the lump sum aid elasticity at $.25$, did not alter the top ranking of foundation aid. Dropping the z aid elasticity to 0 did alter our results. In this extreme case, DMA and FBEA became the preferred policies. By increasing the elasticity of G with respect to z gradually from 0 we can find that elasticity value where FA retains its top ranking. Moving from 0, in increments of $.05$, our simulation results show that for an exogenous aid elasticity of $.1$ or greater (with price and income elasticities of $-.4$ and $.65$) FA is preferred to DMA and FBEA for all objective functions considered here. The $.1$ elasticity is below most recent econometric estimates. See Inman, "The Political Economy." Relaxing other key assumptions also appear to leave the preeminence of lump sum aid unaffected. See Inman, "Optimal Aid," for a full discussion.

policy simulations. Whether such simulations will prove useful in the policy-making process, such as in congressional deliberations over revenue sharing, is another question. The concluding section of this paper offers a few brief comments on this issue.

Policy Models and Aids Policy

The objective of this chapter has been to develop a grants-in-aid policy model for a metropolitan economy which can assist decision makers in the design of a regional aid program. The metropolitan economy is an extraordinarily complex economic system, however, and a policy which appears at first glance to be doing the job may prove in the long run only to make matters worse. A general equilibrium-behavioral model of the effects of grants is therefore needed.

Yet building a policy model which works and then deciding on how that model should be used in the process of social decision making are two quite different issues. Wildavsky[35] and Banfield[36] have argued, for example, that such models are best kept out of the policy process altogether. The models only upset the delicate political balance by undoing hard-won agreements on social values and introducing a potentially powerful but "illegal" player—the unelected "expert." These are real dangers, but I think they can be overcome by introducing the modeling information with care and safeguards.

At a minimum the technical limitations of any policy model must be well understood by legislators or their trusted counselors. Like all models, the one presented here is limited in its scope. First, it is a model of a single urban economy's reaction to grants-in-aid. Fiscal interactions between regions are not considered. Fortunately, when these interregional fiscal effects are unimportant, as available evidence seems to indicate, then a metropolitan model parameterized to each of many regions can be used for state or national policy making. The effects of state or national policies will simply be the sum of the many independent regional effects.

Of more importance to the general applicability of the aids model presented here is our lack of information about the process of local political decision making and the technology of local public services. The model effectively "freezes" the political choice process by assuming a politically stable decision maker before and after the aids policy. This model is

---

[35] A Wildavsky, *The Politics of the Budgetary Process* (Boston: Little, Brown, 1964).

[36] E. Banfield, " 'Economic' Analysis of 'Political' Phenomena: A Political Scientist's Critique." University of Pennsylvania (no date), mimeograph.

probably valid for suburban communities with representative (nonreferendum) governments or large cities with strong political or civil service machines, but it is likely to fail for communities run by referendums or a "competitive" (Downsian) representative process.

The application of the model assumed all communities were equally efficient in the provision of education. Such a strong assumption was necessary because we do not have good information on the technology of educating children. Thus, one of the potentially serious problems of the urban economy, technical inefficiency, could not be analyzed in our simulations.[37] As a result, our conclusions may have been biased away from aid programs, notably FBEA, which tend to promote technical efficiency.

Given its technical accuracy, how should a model such as the one developed here be introduced into the process of choosing policies? Clearly the model cannot stand apart from the choice process; the evaluation rule, W, was an explicit attempt to characterize these deliberations. There are two questions here. First, who should get the modeling information? Second, how should the information be transmitted and assimilated? For representative governments the recipient should be the legislature. The information can be transmitted by the executive branch or by the outside experts who do the modeling, with the most effective channel being a committee or expert-hearing format. The results of the policy modeling can be packaged, as in Table 1, and expanded for a wider range of policy alternatives. Representatives could then debate the merits of the sample policy programs as revealed in summary statistics, select an alternative, or request the executive staff, outside experts, or their own staff to try other alternatives. The emerging policy consensus would reveal an implicit evaluation rule, W, which, if maximized, would have given the compromise policy chosen by the legislature.[38]

The role of the expert and his model in this framework is to advise, not to decide. The United States currently transfers $36 billion annually down-

---

[37] Though, of course, the general model can incorporate these inefficiencies if production function information is available.

[38] A variant of this iterative procedure for revealing decision makers' preferences is found in S. Marglin, *Public Investment Criteria* (Cambridge, Mass.: M.I.T. Press, 1967). B. Weisbrod, "Income Redistribution Effects and Benefit–Cost Analysis," in S. B. Chase, ed., *Problems in Public Expenditure Analysis* (Washington, D.C.: Brookings Institution, 1968); and K. Mera, "Experimental Determination of Relative Marginal Utilities," *Quarterly Journal of Economics*, vol. 83 (August 1969) have in fact estimated W from past legislative decisions. These estimates can be used in subsequent decisions to assist in narrowing the initial search for policies. But as legislatures change, past W's should not be taken as a certain measure of the present preferences.

ward through our federal system.[39] The intention of this sizable resource reallocation is presumably to improve the delivery of public services at the state and local levels. But such reallocations should not be decided in an informational vacuum. Analysis is needed which can be usefully applied in a democratic process of social choice. This chapter has been offered in that spirit.

[39] Census of Governments, *State Government Finances in 1972*, Series GF72–No. 3, (Washington, D.C.: Government Printing Office, 1973), and *City Government Finances in 1971–72*, Series GF72–No. 4, (Washington, D.C.: Government Printing Office, 1973).

5 An Econometric Model of Federal Grants and Local Fiscal Response

MARTIN McGUIRE*

Increasingly, the provision of public goods and services at the state–local level in this country is coming to depend upon a complexity of interacting budgetary decisions made at various levels in our governmental hierarchy. A foremost example is federal grants-in-aid which purposely attempt to influence or manipulate the allocation of local resources in line with federal objectives. As is evident from the literature,[1] economists have directed a

*Professor of Economics, University of Maryland.

The comments of Henry Aaron, Barbara Bergmann, and especially Roger Betancourt are gratefully acknowledged. The research for this chapter has been supported by the Institute of Social and Economic Research, University of York (England); by the Bureau of Business and Economic Research, University of Maryland; and by the National Science Foundation. The original version of this paper, as presented at the Columbus, Ohio meetings of the Metropolitan Governance Research Advisory Committee of Resources for the Future, contained fairly extensive reports of statistical estimation of the theoretical parameters derived in this chapter. Subsequent empirical research by the author has superseded those reported results. The results of work presently underway will be furnished on request as they become available.

[1] In addition to the sources cited in the footnotes which follow, the following articles all illustrate economists' interest in this subject: Thomas E. Borcherding and Robert T. Deacon, "The Demand for the Services of Non-Federal Governments," *American Economic Review*, vol. 62 (December 1972), pp. 891–901; Mark A. Haskell, "Federal Grants in Aid: Their Influence on State and Local Expenditures," *Canadian Journal of Economics and Political Science*, vol. 20 (November 1964); Martin McGuire and Harvey

substantial effort toward investigating the effects of federal grant programs on "local" allocation decisions.[2] With certain notable exceptions, this large body of analysis, however, has failed to distinguish price and income effects of federal grants.[3] The purpose of this chapter, therefore, is to develop empirically testable models for describing the impact of the grant-in-aid system in price and income terms. We are especially interested to examine the effect of grants upon local allocations *among* public sector categories; we are secondarily interested in how grants influence public and private distribution of expenditures. More specifically, we hope to illuminate three interrelated issues bearing upon the effective operation of federal grants.

As in conventional consumer-demand studies we want to examine local responsiveness to price and income changes. Thus one issue to be addressed is, How interchangeable are aided and nonaided outputs in the local decision-makers' preferences? If these goods are close substitutes, and local indifference curves between them are shallow, rather weak federal price incentives can effect a major change in local allocations. If, on the other hand, aided and nonaided outputs are poor substitutes, strong price incentives (or other forms of manipulation) should have to be employed to

---

Garn, "Problems in the Cooperative Allocation of Government Expenditures," *Quarterly Journal of Economics* (February 1969); James C. Ohls and Terence J. Wales, "Supply and Demand for State and Local Services," *Review of Economics and Statistics,* vol. 54 (November 1972), pp. 424–430; Thomas F. Pogue and L. G. Sgontz, "The Effect of Grants-in-Aid on State–Local Spending," *National Tax Journal,* vol. 21 (June 1968); Seymour Sacks and Robert Harris, "The Determinants of State and Local Government Expenditure and Intergovernmental Flows of Funds," *National Tax Journal,* vol. 17 (March 1964); and Lester C. Thurow, "A Theory of Grants-in-Aid," *National Tax Journal,* vol. 19 (December 1966), pp. 373–377.

[2] The term *local* will be used to mean both state and local governments throughout this chapter.

[3] Recent attempts to measure the stimulative effect of federal aid ordinarily show grants to have a much greater impact on local expenditure than does local income. See Jack W. Osman, "The Dual Impact of Federal Aid on State and Local Government Expenditures," *National Tax Journal,* vol. 19 (December 1966); and George Pidot, "A Principle Components Analysis of the Determinants of Local Government Fiscal Patterns," *The Review of Economics and Statistics,* vol. 51, no. 2 (May 1969), pp. 176–188. Such studies have been challenged by Elliot R. Morss and Wallace E. Oates on grounds that federal aid may not be truly independent of state–local expenditures, that aid payments may depend on expenditures, or that each may depend on the other simultaneously. Although such regression models suggest that, statistically, grants are "super-stimulative," they often are not formulated in a way to allow a meaningful economic interpretation of that result. See Morss, "Some Thoughts on the Determinants of State and Local Expenditures," *National Tax Journal,* vol. 19 (March 1966); and Oates, "The Dual Impact of Federal Aid on State and Local Expenditures: A Comment," *National Tax Journal,* vol. 21 (June 1968), pp. 220–223.

change local allocation behavior. Also, we must consider how local decision makers would allocate an income increase among private goods and the various publicly supplied goods.

However, unlike conventional demand studies, we will not have independent information on local incomes and prices, since we do not know in advance in what form grants effectively intervene in local decisions. Thus, a second question, How does the grant-in-aid system alter the resource allocation *opportunities* open to local decision makers? How do grants shift the local budget line in or out, or rotate that budget line? Do grants only alter the effective price paid by localities for the aided function? Or are local governments effectively able to divert or transform categorical grants into revenue-sharing budget supplements?

These above questions pertain essentially to the structure of local decision making and to the impact of grants on such decisions. Another set of issues relates to the allocation strategy of federal grants; that is, to the supply function of grants-in-aid. Hence, our third question is, Do federal grants reward effort by state–local grant recipients (and in the process increase differences in per capita expenditures among states); or do grants equalize local expenditures on the basis of need? The purpose of this chapter is to develop a family of models which (once implemented) will allow answers to these questions to be inferred from statistical study.

The Structure of Federal–Local Allocations and the Consumer-Demand Model

In order to interpret observed resource allocations on the part of various levels of government one requires a model. Hitherto, theoretical studies of this problem have depicted local allocation decisions in the classic model of the individual consumer, that is, as maximization of a preference function subject to resource constraints.[4] Our study also will employ a consumer-allocation model, but the classic model will be modified to reflect the peculiarities of the grant system, thus permitting an empirical estimation of how that system operates. Before formulating the model, however, we must address two questions which call into doubt the validity of using it at all.

[4] See, for example, four articles by Edward M. Gramlich, "Alternative Federal Policies for Stimulating State and Local Expenditures: A Comparison of Their Effects," *National Tax Journal*, vol. 21 (June 1968); "State and Local Governments and Their Budget Constraints," *International Economic Review* (June 1969), pp. 163–182; "The Effect of Federal Grants on State–Local Expenditures: A Review of the Econometric Literature," paper presented to the National Tax Association, September 1969; and with Harvey Galper, "State and Local Fiscal Behavior and Federal Grant Policy," *Brookings Papers on Economic Activity*, vol. 1 (1973), pp. 15–58.

First, there is doubt that complicated bureaucratic decisions can be represented as conscious, utility-maximization processes. For our purposes we can dispense with the conscious maximization hypothesis, provided we assume that local bureaucracies show consistent responses to price and budget changes.[5] The essential fact that local decisions depend on both preferences and opportunities simultaneously does not necessarily mean utility maximization is the only or best model of local behavior. Other norms such as budget- or bureau-maximizing practices should be explored; the empirical method presented in this study should be of use in making such comparisons.[6] A second, more serious problem is whether grants-in-aid can validly be characterized as price-plus-income changes with the implicit assumption that local governments retain freedom to make marginal allocation choices once grants are introduced into the process. I want to consider this question at length: first, to clarify the behavioral context in which the consumer model can be applied, and second, to argue on theoretical grounds that such a context may well, in fact, obtain.

For some very simple aid mechanisms a grant merely changes price and income. In Fig. 1, for example, if line *B* shows the local budget before federal grants, line *C* can be used to represent an unconditional grant or a general increase in local resources and therefore a pure income effect. Line *D* represents an open-ended but conditional (or matching) grant as a pure price change; lines *E* and *F* show combinations of simple price and income changes. If the grant system functions in any one of these ways, one can validly explain local allocation choices (assumed to be point *a* in Fig. 1) in terms of income and prices. Other, more complex grant mechanisms, however, may confront a community not with continuous marginal choices but with a discrete take-it-or-leave-it offer. Figure 2 offers one such example, a fixed-size categorical grant which requires no matching local funds, but is policed to ensure no reduction in local effort. This device also moves the community from point *b* to *a*, although point *a* is not caused by price–income changes. Note, however, that point *a* would have been the outcome under a marginal choice process for any of an infinite number of *multipart pricing* offers by federal authorities. For example, a fixed charge of *ed* followed by a price subsidy (shown by line *dca*) would equally have stimulated selection of point *a*. In this case, income and price effects of the

---

[5] In the long run, federal intervention may change local attitudes and perceptions of needs as well as the responsiveness of local officials to client demands. These effects are not covered in this chapter.

[6] For theoretical investigation along this line, see William Niskanen, "The Peculiar Economics of Bureaucracy," *American Economic Review Papers and Proceedings* (May 1969). Also see Martin McGuire, "Cost Versus Performance Subsidies as Tools of Intergovernment Finance," *National Tax Journal* (March 1971).

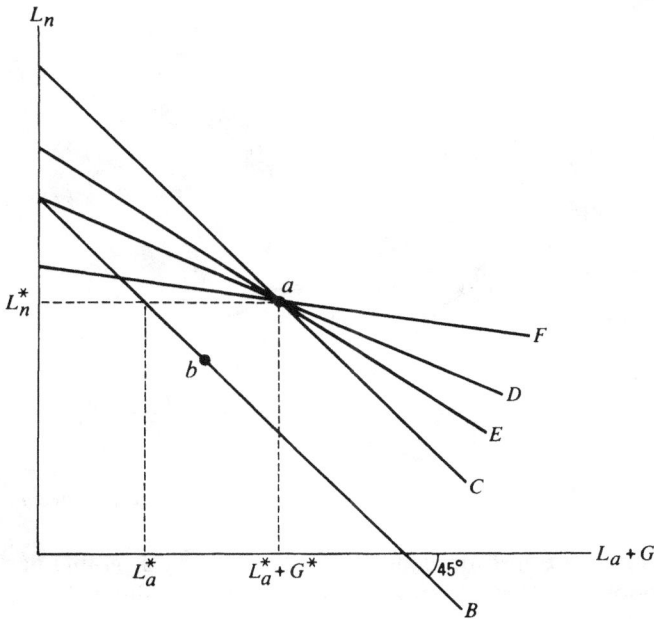

FIG. 1. Price and income grants.

federal grant could be identified with the first and second parts, respectively, of the multipart price. (Another multipart price with the same result is price *ec* up to point *c*, followed by price *ca* thereafter.)

Which account is more realistic for describing grant systems? There is probably no single answer to this question, since procedures vary among different functional programs and levels of government. Although we cannot settle the question conclusively, a strong argument can be made in favor of the multipart price, continuous-choice alternative. For example, the grant-in-aid system is embedded in a political process with a large element of bargaining and negotiation; hence there are built-in incentives for striking mutually beneficial bargains. In particular, key provisions of grant programs—such as matching ratios and budget allotments to states and other jurisdictions—are not determined arbitrarily. Rather, they are agreed upon in a process of legislative–executive bargaining. Further, the very intricacy of rules and regulations for grant controls makes them vulnerable to manipulation by local governments. And finally, in many cases local authorities may have perfectly legal options available for defeating the nominal purpose of grants.

Consider, for example, the variety of options which a local authority

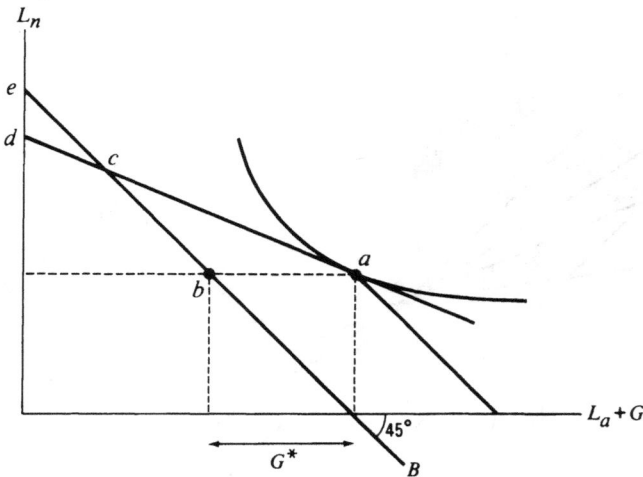

FIG. 2. The equivalence between mandated local expenditures and price–income grants.

might exercise to transform conditional, or categorical, grants back into fungible resources. One opportunity to thwart the conditional purpose of a federal subsidy will arise when local leaders can trade the subsidized good on a "market"; for by so doing, they will reconvert a conditional subsidy to a general income supplement. Sales of the services of a subsidized public facility (say, water or waste treatment) to another local authority will accomplish this purpose. If direct sales of the subsidized output are physically impossible or legally prohibited, then other indirect forms of exchange may be possible: the local government may possess an equivalent, salable good, or local authorities may convert conditional to unconditional resources—not by selling more or less substitutable goods, but by renting them out or imposing user charges. Alternatively, categorical grants may in effect be traded for general resources through time; where a local government plans to invest in future public capital projects, it may borrow money for present construction, thereby using the federal cost share to reduce future tax loads. Or again, when the benefits of a public facility appear in the form of increased profits to local firms or increased incomes to local citizens, the local government will recover a portion of the subsidy as local taxes. Possibly the greatest opportunity for defeating intended conditional effects occurs in cases where grants are supposed to apply only to increases in local output over current levels. By understating or reducing normal funding to the subsidized programs, by using a project which would be undertaken in any case as the vehicle for securing a matching grant, by redefining budget categories, or by a judicious allocation of overhead costs, local officials may, in effect, convert the grant to a pure income supplement. To the extent that such options are exercised by local government, the nominal legal provisions of grant programs will not reveal the actual change in local opportunities caused by federal intervention, and as a theoretical construct the traditional consumer model can plausibly be transplanted into the context of a local government allocation.

In view of local options for converting conditional to fungible resources, the way to discover the *effective* change in the resource constraint induced by a federal grant is probably not to sift through the morass of federal rules and regulations concerning matching ratios and enforcement procedures. Obviously, reliable prior information on the operation of grants should not be discarded, but even when such information is available, determination of the changes in local opportunities induced will remain largely a statistical problem. In view of the foregoing arguments, we shall proceed on the assumption that price and income effects can explain local allocations. Given this assumption we will show how to estimate the federal multipart pricing system most likely to have caused the observed allocations. At the very least, the consumer-allocation model must be regarded as a possibility, and empirical tests of that income–price model as a limiting bench mark for interpreting the data.

A Model for Simultaneous Determination of the Effect of Grants on Local Resource Constraint and Allocation Choices

Theoretical Formulation of the Price and Income Components of Grants

Figure 1 is useful for illustrating a structural model in which the effective price and income changes induced by federal grants are unknown parameters of the grant-in-aid system—parameters to be estimated by empirical analysis.[7] We shall carry the analysis forward for two categories of local spending, one aided by a higher level of government, and the other not aided.[8] Therefore we divide the expenditures of local government into three categories: L_n, representing expenditures on nonaided functions; L_a, repre-

---

[7] Throughout this chapter *price* refers not to the various real unit costs which differ among regions and jurisdictions but rather to the price change brought about by a grant as perceived by local officials. We have ignored real cost differences because they are quantitatively small compared to grant rates.

[8] Allowing for more than a single aided category requires a multiequation model. Such simultaneous analysis of diverse grant programs for diverse functional categories will not be attempted. We can use the analytics developed in this to study one aided category at a time, or to study the average of all aided categories taken together (although the problems of such aggregation should not be minimized).

senting expenditures on the aided function (*where both these expenditures are from local resources*); and G, representing grants received from other levels of government to supplement the aided function. (We shall use L_a, L_n, and G to denote dollar expenditures; Q_a and Q_n to denote physical quantities or indices thereof; and p to denote the price of Q_a.) Time-series or cross-sectional data will show a number of observations of realized outcomes for the various local decision units. Figure 1 shows one such observation as a point labeled a. This point implies a local expenditure L_a^* on aided categories, L_n^* on other categories, and a grant of G^*. Line B (a 45-degree line) shows how the total observed local government expenditures ($L_a^* + L_n^*$) might feasibly have been allocated otherwise.

The basic assumption being that the particular combination a was *chosen* by the local government, the question naturally arises as to which budget limitation actually constrained the observed choice. Figure 1 shows four possibilities, all consistent with the observed grant, G^*. Corresponding to each of these alternative budget constraints is a different set of local preferences explaining the selection of point a.

Line C represents a pure budget-supplementing grant with no price effect; line D, a pure price subsidy with no budget effect; line E, a two-part price consisting of budget supplement plus a price subsidy; and line F, a two-part price consisting of an effective budget reduction plus an overcompensating flat price subsidy.

Since the price and income components of observed grants are unknown, we are ignorant of what combination of the above constraints actually restricted the local decision. Thus we want to formulate federal–local allocations in such a way as to allow statistical tests to reduce this ignorance.

Assuming federal intervention to take the effective form of multipart pricing, the local government will find a portion of the grant to be a fixed cost or revenue supplement. The remainder of the grant is devoted to designated categorical purposes, in accord with some *effective* matching requirement (in contrast to the nominal or legal matching ratio).[9] Figure 3 shows the model in detail for the particular hypothetical observation (L_n^*, L_a^*, G^*). We postulate that G_1, a part of the total grant, is in effect a pure revenue-sharing budget supplement (G_1 may be positive or negative); G_1 is the fixed portion of the multipart price. Total fungible resources available to local

---

[9] One can imagine various processes with this result: for example, federal grants purposely might be part categorical, part revenue sharing; federal enforcement procedures might allow a portion of grants to be diverted as revenue supplements; or if federal and local representatives in effect bargain simultaneously for the size of federal grant allotment and for an ex post matching ratio, the effect is equivalent to a multipart pricing system. In any case, the net effect of local options for defeating matching requirements will be reflected in this derived price ratio.

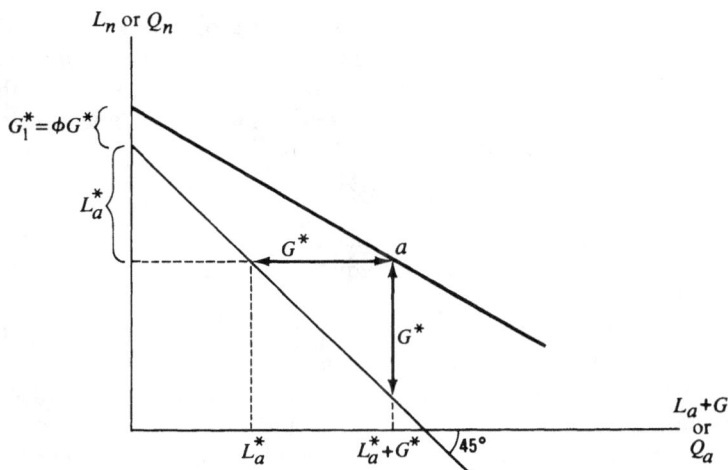

FIG. 3. The decomposition of a grant into effective price and income components.

officials, that is, the total local government budget B (including G_1) therefore is

(1) $$B^* = L_a^* + L_n^* + G_1^*.$$

In selecting point a, the local government actually chooses to expend $L_a^* + G_1^*$ of this total budget on Q_a,[10] paying an effective price of

(2) $$p^* = \frac{L_a^* + G_1^*}{L_a^* + G^*} = 1 + \frac{G_1^* - G^*}{L_a + G^*}$$

(equal to the slope of its postgrant budget line). The variable p therefore represents the postsubsidy price of the aided category as a percentage of total unit cost. While local officials spend $(L_a^* + G_1^*)$ of their fungible resources on the aided function, the quantity they receive costs $(L_a^* + G^*)$. Thus we have broken down the total grant-in-aid into two component parts, G_1, an income component, and $(G_1 - G)$, a price change component.[11]

[10] One could break down the local expenditure decision on Q_a further as follows: first, the local government might be assumed to divide G_1 between Q_a and Q_n in the proportions $\alpha_1 : 1 - \alpha_1$; then, to comply with the federal requirement that all of G_1 actually show in the books as allocated to the aided function, $G_1(1 - \alpha_1)$ is transferred out of category Q_n into Q_a, and equal amount of *local* funds are transferred from Q_a to Q_n.

[11] The diagram used to illustrate this breakdown is based on the assumption that federal grants change only the division of the local budget between L_a and L_n, not the total budget $(L_n + L_a)$ itself. It is to be emphasized that this is an assumption made for expository purposes at this stage of the argument. Tax reduction effects will be incorporated into the model presently. (Clearly our formulation of the incomes–price components of a grant will not be changed by the introduction of tax–budget effects.)

Although this formulation theoretically allows one to separate price from income effects of grants, the problem remains that neither component is directly observable. Suppose, however, we were willing to assume that the federal grant policy works in such a way that G_1 is some constant proportion of G, over all observations; that is,

$$(3) \qquad\qquad G_1 = \phi G.$$

It might be argued that it is inconsistent to assume that G_1 is a fixed proportion of G and at the same time equivalent to a fixed, unconditional income supplement. But although such an hypothesis would be inconsistent for a single individual who foresaw, understood, and responded to the entire federal grant package at once, it is in no way inconsistent for a decision group. Moreover, the hypothesis is only that the federal-local system works *as if* it consisted of a pure income plus a pure price component. Other political and administrative processes may be observationally equivalent to this hypothesis.[12] Accordingly, one interpretation of equation 3 should be that federal enforcement allows a constant proportion of grants to "leak" into the revenue-sharing category if ϕ is positive, or that the federal government exercises a degree of monopoly power through the multipart price if ϕ is negative.

Equation 3 implies a local budget constraint of

$$(4) \qquad\qquad B = L_a + L_n + \phi G$$

and a local price for the aided function of

$$(5) \qquad\qquad p = \frac{L_a + \phi G}{L_a + G} = 1 + \frac{(\phi - 1)G}{L_a + G} = 1 + (\phi - 1)q$$

both of which depend on observed data plus an unknown parameter.[13] As we shall now demonstrate, this hypothesis can be subjected to econometric test and the parameter ϕ can be estimated.

Econometric Formulation of the Local Allocation Decision

We can now specify the local decision process in one of two ways. First, we can directly postulate a local "demand function." When data on prices, quantities, and budgets are all available, the analyst ordinarily has wide choice as to alternative functional forms for such a relationship. In our problem, however, prices and budgets are not given as data. The parameter ϕ, which determines the division of a grant between price and income com-

---

[12] See Arthur S. Goldberger, *Econometric Theory* (New York: John Wiley, 1964), pp. 306–310, for a discussion of observationally equivalent structures.

[13] We define $q \equiv \dfrac{G}{L_a + G}$.

ponents, is unknown. Consequently, only certain functional forms permit linear estimation techniques to identify the key parameters of interest. More specifically, this feature of our problem requires that local *expenditures* be taken to depend on a polynomial in budget and price. For example, as will be demonstrated, the linear expenditure function,

(6)   [Expenditure on Aided Outputs] $= \alpha_0 + \alpha_1(B) + \alpha_2(p)$

is identifiable, whereas theoretically more desirable logarithmic or exponential forms with constant price elasticities are not.

An alternative approach to formulating the local allocation decision is to derive a demand function from a classic utility-maximization process. (The demand function derived in this way would have to be a polynomial.) The fact that we do not assume local bureaucracies consciously to maximize the representative (or median[14]) voter's preference function might argue for simply postulating a demand function. Moreover, imposing a utility-maximization process on the data may impute an implausible degree of "consistency" to local decisions. I have chosen this latter alternative, nevertheless, because of the advantage of working in the context of the established body of demand theory. The various implicit utility functions considered, therefore, should be regarded as revealing the implicit preferences of the bureaucrats and not necessarily those of the people.

A linear expenditure system: Disregarding tax relief effects of grants. One utility function which will allow us to separate out and estimate the price- and income-changing components of a grant is the *Stone–Geary* form. Initially, let us disregard possible tax relief effects of grants, and assume that local officials maximize

(7) $U(Q_n, Q_a) = (1 - \alpha_1) \log \left(Q_n + \dfrac{\alpha_0}{\alpha_1} \right) + \alpha_1 \log \left(Q_a - \dfrac{\alpha_2}{1 - \alpha_1} \right)$

subject to $B = Q_n + pQ_a$.

The variables Q_a and Q_n indicate physical quantities of aided and nonaided outputs; B indicates the local budget including the straight income component of the grant received; p indicates the price of aided outputs including the effect of grants, as was shown in equation 2; and the price of Q_n is assumed to be unity. In this formulation, p is taken to be a param-

---

[14] On the allocation behavior of a government which attempts to maximize the preferences of a (shifting) median voter, see David Bradford and Wallace Oates, "Towards a Predictive Theory of Intergovernmental Grants," *American Economic Review Papers and Proceedings* (May 1971), pp. 440–448.

eter (its value as yet unknown to us). First-order conditions for a maximum follow from differentiating with respect to Q_a and Q_n:

(8)
$$\frac{p}{(Q_n + \alpha_0/\alpha_1)} = \frac{\alpha_1/(1 - \alpha_1)}{Q_a - \alpha_2/(1 - \alpha_1)}$$

or

$$pQ_a = \frac{\alpha_0}{1 - \alpha_1} + \frac{\alpha_1}{1 - \alpha_1} Q_n + \frac{\alpha_2}{1 - \alpha_1} p.$$

One more manipulation of this equation shows expenditures to be a linear function of price and local budget.

(9)
$$pQ_a = \alpha_0 + \alpha_1 B + \alpha_2 p.$$

Stone's linear expenditure system has been widely investigated, and its properties are well known.[15] The parameters $-\alpha_0/\alpha_1$ and $\alpha_2/(1 - \alpha_1)$, for example, are taken as "subsistence levels" of consumption of Q_n and Q_a, respectively. All goods are gross complements, but the utility-compensated, cross-price effect is positive $(\partial q_i/\partial p_j)\big|_{U=\text{constant}} > 0$, so that all goods are specific substitutes. The own-price elasticity of demand must lie between 0 and -1 so that demand is inelastic, so long as "subsistence levels" of consumption of Q_n and Q_a are positive. Otherwise elasticity values below -1 are possible.

If we define a unit of Q_a as a physical quantity with a total cost of $1.00, then $Q_a = L_a + G$,[16] and in place of equation 9 one may write

(10)
$$L_a + G_1 = \alpha_0 + \alpha_1[L_n + L_a + G_1] + \alpha_2\left[\frac{L_a + G_1}{L_a + G}\right]$$

as the structural relation to be estimated. Substituting $G_1 = \phi G$, the structural equation then becomes

(11)
$$L_a + \phi G = \alpha_0 + \alpha_1(L_n + L_a) + \alpha_1\phi G + \alpha_2\left[\frac{L_a + \phi G}{L_a + G}\right].$$

To transform this equation to an estimable form, we move ϕG to the right and $\alpha_1 L_a$ to the left of the equality sign, and alter the price variable

---

[15] See Arthur S. Goldberger, "Functional Form and Utility: A Review of Consumer Demand Theory," Systems Formulation, Methodology, and Policy Workshop paper 6703, Social Systems Research Institute, University of Wisconsin (October 1967).

[16] It is assumed that local expenditures are honestly reported, so that grant expenditures were in fact made in the reported categories.

according to equation 5: $p = 1 + (\phi - 1)q$;

$$q \equiv \frac{G}{L_a + G} .^{17}$$

The equation to be estimated thereby becomes,[18]

$$(12) \quad L_a = \frac{\alpha_0 + \alpha_2}{1 - \alpha_1} + \frac{\alpha_1}{1 - \alpha_1} L_n + \frac{(\alpha_1 - 1)\phi}{1 - \alpha_1} G + \frac{\alpha_2(\phi - 1)}{1 - \alpha_1} q + u$$

where u represents error term.

Tax relief effects of grants. The preceding formulation of the local allocation decision does not explicitly allow for the possibility that local taxes may be affected by grants although some correlation between local contributions $(L_a + L_n)$ and the size of the grant (ϕG) may be due to this cause. To incorporate this possibility into the analysis, we will assume that only a fraction of $G_1 = \phi G$ actually augments the local public budget. Let $(1 - \pi)$ of the revenue-sharing (or fixed cost) portion of grants be given over to increased private expenditure via tax relief.[19] This assumption, as illustrated by Fig. 4, changes the structural equation to

$$(13) \quad L_a + \phi G = \alpha_0 + \alpha_1(L_n + L_a + \pi\phi G) + \alpha_2 \left[\frac{L_a + \phi G}{L_a + G} \right]$$

and the reduced form equation to

$$(14) \quad L_a = \frac{\alpha_0 + \alpha_2}{1 - \alpha_1} + \frac{\alpha_1}{1 - \alpha_1} L_n + \frac{(\alpha_1\pi - 1)}{1 - \alpha_1} G + \frac{\alpha_2(\phi - 1)}{1 - \alpha_1} q.$$

[17] It should be emphasized that implicit in this formulation is the assumption that different states may face different effective matching ratios, that is, values of q.

[18] Evidently to estimate the structural relation (equation 11) by "reduced" form (equation 12) one must handle the problem that $q \equiv G/(L_a + G)$ is negatively correlated with the error term; hence the estimated coefficients will be biased. The direction of bias on the price coefficient is definitely negative, for the L_n-coefficient probably negative, and for the G-coefficient probably positive.

[19] This formulation implicitly assumes first, that the total tax-expenditure decision is independent of (presumably prior to) the expenditure *mix* decision; second, that local taxes equal $(L_a + L_n)$, which ignores local debt financing of expenditures; and third, that a direct income transfer $(T = \phi G)$ causes a reduction in local outlays according to

$$\frac{d(L_a + L_n)}{dT} = -(1 - \pi).$$

Thus the observed effect upon L_a of a change in $T = \phi G$ is

$$\frac{dL_a}{dT} = \frac{\partial L_a}{\partial(L_a + L_n + T)} \left[\frac{d(L_a + L_n + T)}{dT} \right]$$

or

$$\alpha_1\pi = \alpha_1[-(1 - \pi) + 1].$$

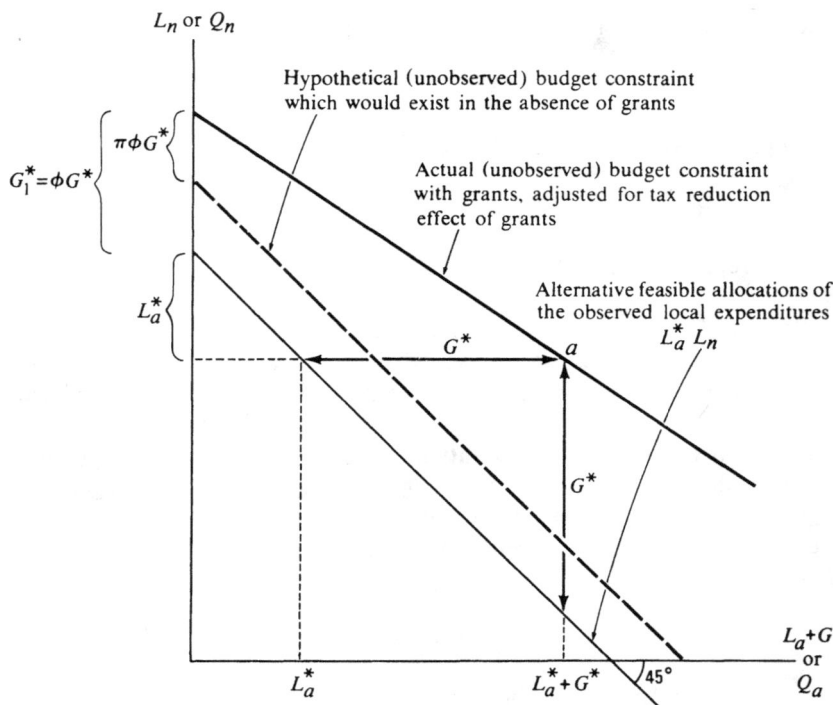

FIG. 4. The decomposition of a grant into tax relief, price, and income components.

Entering the tax relief-effect parameter π in this way allows us to estimate the tax relief effect of grants without assuming a grant to have the same impact on taxation as does internal private income growth. But, in equation 12, the parameter α_1 is unambiguously identified, while the other parameters are not. One must assume a value for one of the four parameters α_0, α_2, ϕ, or π to solve for the remaining three. These parameters have a behavioral interpretation as follows:

(15) $\quad -\dfrac{\alpha_0}{\alpha_1}, \dfrac{\alpha_2}{1-\alpha_1} = $ "subsistence" quantities of Q_n and Q_a.

where

ϕ = budget-supplementing, or revenue-sharing fraction of the total grants.

π = proportion of the "revenue-sharing" fraction of grants which local governments allocate to their public budget. We expect $0 < \pi < 1$ provided neither public nor private goods are "inferior."

α_1 = marginal propensity to expend fungible public resources on the aided functions. We expect $0 < \alpha_1 < 1$.

$\epsilon_y = \alpha_1\left(\dfrac{L_n + L_a + \phi G}{L_a + \phi G}\right)$ "local budget elasticity" of local demand for aided outputs.[20]

α_2 = price sensitivity of local resource allocations to the aided functions.

$\epsilon_p = \dfrac{\alpha_2}{L_a + G} - 1$ = price elasticity of local demand for aided outputs.[21]

Formulating the local decision process in this way will allow statistical techniques simultaneously to estimate how grants influence resource constraints and how those constraints, in conjunction with preferences, determine local choices. Our general statistical formulation will allow for three "pure" cases, and, of course, intermediate outcomes.

Absence of income and price effects. Legal formulas for state–local efforts notwithstanding, if federal grants have no influence at all on local decisions with regard to mix or level of budget, then federal funds for specified purposes will increase total outlays for those purposes by the exact amount of the federal funds. This conclusion would be suggested by parameter values: $\phi = 1$; $\alpha_2 = 0$; $\pi = 1/\alpha_1$.

Pure income effect. If all federal funds were absorbed in local budgets and treated as completely fungible with local internal resources, the structural parameters would be: $\alpha_2 = 0$; $\phi = 1$; $\pi = 1$.

Pure price effect. Third, local decision makers may view grants as pure price changes. This would be reflected by coefficients: $\phi = 0$. Before the

---

[20] If Q_a denotes an amount whose total real cost is $\$1.00$, then $\alpha_1 = p\Delta Q_a/\Delta B$; where B = local budget. Therefore:

$$\epsilon_y = \frac{\Delta Q_a/Q_a}{\Delta B/B} = \frac{p\Delta Q_a/pQ_a}{\Delta B/B} = \frac{B\alpha_1}{pQ_a}$$

$$= \alpha_1\left[\frac{L_n + L_a + \phi G}{L_a + \phi G}\right].$$

[21] Using the same notation as in footnote 20:

$$\alpha_2 = \frac{\Delta(pQ_a)}{\Delta p} \frac{p\Delta Q_a + Q_a\Delta p}{\Delta p}$$

and

$$\frac{\alpha_2}{Q_a} - 1 = \frac{p\Delta Q_a}{Q_a\Delta p} = \epsilon_p.$$

But $Q_a = L_a + G$, since Q_a is defined with a unit cost of $\$1.00$.

evidence is in, it seems reasonable to expect none of these pure hypotheses to be supported by the data and instead some intermediate case to arise.[22]

Introduction of preference-normalizing variables. One could proceed directly to test equation 14 against data if one thought that differences among the various observations were derived only from price–income effects and not from variations in the underlying preference function. Tests of our model, however, are most likely to involve data, whether time-series or cross section, from diverse governments with different preferences. In addition to budget and price variables as explicitly developed in previous sections, therefore, we shall use certain other variables as indicators of "local needs" for the purpose of washing out differences in the preference structures of different states. Thus we hope to normalize the effects of price and budget on a single, uniform preference structure across states. Our preference-normalizing variables could interact with price and resource variables in various ways. One alternative is merely to add such extra variables to equation 14. Using N_i to designate need variables, the equation to estimate, in this case becomes

---

[22] Inasmuch as selection of the Stone–Geary utility form has been somewhat arbitrary, it should be emphasized that other functional forms also produce identifiable demand structures.

It turns out, for example, that a generalized quadratic utility function, although not usually estimable, is, so far, our special problem. The constrained maximization problem is

$$\text{Max } \alpha_0 Q_a - \frac{\alpha_1}{2} Q_a^2 + \alpha_2 Q_n - \frac{\alpha_3}{2} Q_n^2 + \alpha_4 Q_n Q_a$$

subject to $B = Q_n + pQ_a$.

First-order conditions require

$$\alpha_0 - \alpha_1 Q_a + \alpha_4 Q_n = p(\alpha_2 - \alpha_3 Q_n + \alpha_4 Q_a);$$

substituting $L_a + G = Q_a$ and $(L_a + G_1)/(L_a + G) = p$, one obtains

$$L_a = \frac{\alpha_0}{\alpha_1 + \alpha_4} + \frac{\alpha_3 + \alpha_4}{\alpha_1 + \alpha_4} L_n - \frac{\alpha_1}{\alpha_1 + \alpha_4} G - \frac{\alpha_4}{\alpha_1 + \alpha_4} G_1 + \frac{\alpha_3}{\alpha_1 + \alpha_4} pL_n - \frac{\alpha_2}{\alpha_1 + \alpha_4} p.$$

Substituting, in turn, $G_1 = \phi G$ and

$$p = 1 + \frac{(\phi - 1)G}{G + L_a} = 1 + (\phi - 1)q$$

one obtains as a reduced-form equation:

$$L_a = \frac{\alpha_0 - \alpha_2}{\alpha_1 + \alpha_4} + \frac{\alpha_3 + \alpha_4}{\alpha_1 + \alpha_4} L_n - \frac{\alpha_1 + \phi\alpha_4}{\alpha_1 + \alpha_4} G + \frac{\alpha_3(\phi - 1)}{\alpha_1 + \alpha_4} L_n q - \frac{\alpha_2(\phi - 1)}{\alpha_1 + \alpha_4} q$$

with five regression coefficients and six structural parameters α_0, α_1, α_2, α_3, α_4, and ϕ. Any of the first five parameters, however, may be given an arbitrary value.

$$(16) \quad L_a = \frac{\alpha_0 + \alpha_2}{1 - \alpha_1} + \frac{\alpha_1}{1 - \alpha_1} L_n + \frac{(\alpha_1\pi - 1)\phi}{1 - \alpha_1} G$$

$$+ \frac{\alpha_2(\phi - 1)}{1 - \alpha_1} q + \frac{\Sigma\alpha_i N_i}{1 - \alpha_1} + u.$$

In this case one is not necessarily interested so much in the coefficients on the extra need variables themselves, as in their influence on the estimates of the structural parameters. In the context of a Stone–Geary expenditure system, a second alternative is to incorporate the need variables in the minimum-survival bundle. The underlying preference function now is taken as

$$(17) \quad U = (1 - \alpha_1) \log \left[Q_n - \frac{\Sigma\alpha_i^n N_i - \alpha_0}{\alpha_1} \right]$$

$$+ \alpha_1 \log \left[Q_a - \frac{\Sigma\alpha_i^a N_i + \alpha_2}{1 - \alpha_1} \right].$$

We maximize this preference function subject to a budget constraint, and then substitute price and tax relief effects; the result is the following equation:

$$(18) \quad L_a = \frac{\alpha_0 + \alpha_2}{1 - \alpha_1} + \frac{\alpha_1}{1 - \alpha_1} L_n + \frac{(\alpha_1\pi - 1)\phi}{1 - \alpha_1} G + \frac{\alpha_2(\phi - 1)}{1 - \alpha_1} q$$

$$+ \Sigma \left[\frac{\alpha_i^a - \alpha_i^n}{1 - \alpha_1} N_i \right] + \Sigma \left[\frac{\alpha_i^a(\phi - 1)}{1 - \alpha_1} N_i q \right] + u$$

where $\quad q = \dfrac{G}{L_a + G}$.

When preference-normalizing variables are incorporated in the Stone-Geary expenditure system as elements in the subsistence bundles, the new parameters α_i^a and α_i^n are of interest, first, because they specify subsistence consumption, and second, because they influence one's estimate of price elasticity. The expressions for these terms are:

$$\frac{\Sigma\alpha_i^n N_i - \alpha_0}{\alpha_1} = \text{subsistence consumption of } Q_n$$

$$\frac{\Sigma\alpha_i^a N_i + \alpha_2}{1 - \alpha_1} = \text{subsistence consumption of } Q_a$$

$$\frac{\Sigma\alpha_i^a N_i + \alpha_2}{L_a + G} - 1 = \epsilon_p, \text{ price elasticity of demand.}$$

Otherwise, structural parameters have the same interpretation as shown on pages 128–129.

In this section, we have fashioned an approach to government allocation which is parallel in large part to the analysis of the individual consumer decision. In essence, we have asked, Assuming the observed pattern of local government expenditures to result from a free, utility-maximizing choice by local officials, what combination of budget constraint and preference pattern (adjusted for differences in need) and federal strategy with regard to allocation of grants best explains the observations?

Extensions of the Basic Model and Procedures for Analyzing Price and Income Components of Grants

In framing this question and building a model to answer it, we are constructing one research tool for application in comprehensive systems analyses of the governmental allocation process. These processes themselves, however, may be quite heterogeneous, so that no single model can capture the essential features of more than a few grant programs. For instance, the analyst may want to discern finer differences than allowed by our basic distinction between price and income components of grants. Accordingly, this section shows how finer distinctions can be incorporated into the basic model. The primary purpose of this section, therefore, remains methodological: to identify alternative structural hypotheses, to demonstrate the identifiability of these structures, and to suggest the interpretation of corresponding regression analyses. The regression models to be developed are as follows:

Model A examines the hypothesis that grants have only an income effect on local allocation decisions, by assuming a priori that grants-in-aid are unconditional budget supplements. This model also assumes that grants are predetermined independently by the federal government.

Model B examines the hypothesis that grants bring about both price and income changes, as shown in equations 14, 16, and 18. Expenditure on aided categories is a linear function of price and budget. As in Model A, grants are taken to be independently determined variables.

Model C examines the hypothesis that the federal government discriminates in its grant-pricing administration among grant-receiving communities on the basis of the size of grant received.

Models D and E examine the hypothesis that the supply of grants, rather than being exogenous as in Models A through C, depends upon local outlays in accord with some effort or equity criterion.

Model A— Determining Whether Grants Have Income Effects Only

If grants were, in effect, merely revenue-sharing instruments and therefore induced no price changes, one should expect a $1.00 increase in grants to call forth π times the increment in local expenditure on aided functions than a $1.00 increase in local public resources would generate. This is equivalent to an a priori assumption that $\phi = 1$ in the price–income equations 11 through 18. Substitution of $\phi = 1$ in that model changes the structural equation to

$$(19) \quad L_a + G = \alpha_0 + \alpha_2 + \alpha_1(L_n + L_a + \pi G) + (\alpha_s^a - \alpha_s^n)S$$
$$+ (\alpha_y^a - \alpha_y^n)Y + (\alpha_u^a - \alpha_u^n)U + (\alpha_d^a - \alpha_d^n)D$$

with a reduced form: $L_a = \beta_0 + \beta_1 L_n + \beta_2 G + \beta_s S + \beta_y Y + \beta_u U + \beta_d D$. If $\pi = 1$ is assumed, one should expect $\beta_2 = -1$.[23] Alternatively π, if calculated, should be in the range 0 to 1. Table 1 shows identification for this simple model.

Model A is representative of most current and historical analysis of federal grants. As summarized by Gramlich (1969) most of these studies show grants to increase local expenditures (and hence in the long run, local taxes as well). Analysis of Model A therefore will probably raise the question of what mechanisms other than a pure income effect might account for the superstimulative or "completely additive"[24] effect of grants. One obvious alternative hypothesis is that grants encourage substitution among functions at the local level by altering effective prices. Another is that the observed data *could* be generated not by any local choice at all, but rather by very specific, highly policed federal grants. In this case, the difference $(1.0 - \alpha_1\pi)$ would represent a leakage or diversion of specific federal funds back into the fungible category. This is an unlikely explanation in view of the variety of options which local authorities have for circumventing the categorical intent of federal grants, and I believe the hypothesis should be rejected. As another possible explanation, the observed data may stem simultaneously from a pure income effect on the local resource constraint in combination with an independent, federal grant-allocation strategy. Later, we shall incorporate independent federal behavior with regard to allocating grants as an additional structural equation, in Models D and E.

Model B—Incorporating Price and Income Effects

If results from Model A suggest that grants for local expenditures have more than a mere income effect on local governments, and if a plausible

---

[23] Thus, $\beta_2 = [(\pi\alpha_1 - 1)/(1 - \alpha_1)]$. Therefore, $\pi = 1$ implies $\beta_2 = -1$.

[24] A grant is said to be "completely additive" to local expenditures when total expenditures on the function increase only by the amount of the grant.

TABLE 1. Identification of Model A

Structural equation:	$L_a + G = \alpha_0 + \alpha_2 + \alpha_1 L_n + \alpha_1 L_a + \alpha_1 \pi G + (\alpha_s{}^a - \alpha_s{}^n)S$
	$\qquad + (\alpha_y{}^a - \alpha_y{}^n)Y + (\alpha_u{}^a - \alpha_u{}^n)U + (\alpha_d{}^a - \alpha_d{}^n)D$
Reduced form:	$L_a = \beta_0 + \beta_1 L_n + \beta_2 G + \beta_s S + \beta_y Y + \beta_u U + \beta_d D$
Identification:	$\alpha_1 = \beta_1/(1 + \beta_1); \qquad \alpha_0 + \alpha_2 = \beta_0/(1 + \beta_1)$
	$\pi = (1 + \beta_1 + \beta_2)/\beta_1$

explanation for the highly "additive" character of grants is that grants in effect alter the *price* local decision makers confront for the aided categories of expenditure, then this price-changing effect should be reflected in the model that one tests.

It will be recalled that Model B, or the basic model developed previously, has the structural form

(20) $\quad L_a + \phi G = \alpha_0 + \alpha_1 L_a + \alpha_1 L_n + \alpha_1 \pi \phi G + \alpha_2 p - \Sigma \alpha_i{}^n N_i + \Sigma \alpha_i{}^a N_i p$

where

$$p = \text{price} = \frac{L_a + \phi G}{L_a + G} = \frac{(\phi - 1)G}{L_a + G} + 1 = (\phi - 1)q + 1.$$

(This model, it should be noted, assumes ϕ is a constant across all communities.) Derived structural parameters for Model B are shown in Table 2. Note that the parameters π, ϕ, α_0, and α_2 are underidentified. In a sense, therefore, Model B is not strictly comparable with Model A (where π was unambiguously identified). We have handled this problem by expressing the various structural parameters as functions of π. Accordingly, estimation of Model B will allow one to infer both the degree to which grants change

TABLE 2. Identification of Model B

Structural equation:	$L_a + \phi G = \alpha_0 + \alpha_1 L_a + \alpha_1 L_n + \alpha_1 \pi \phi G + \alpha_2 p + \Sigma \alpha_i{}^a N_i p - \Sigma \alpha_i{}^n N_i$
Reduced form:	$L_a = \beta_0 + \beta_1 L_n + \beta_2 G + \beta_3 q + \beta_s S + \beta_y Y + \beta_u U + \beta_d D$
	$\qquad + \beta_{sq} Sq + \beta_{yq} Yq + \beta_{uq} Uq + \beta_{dq} Dq$
	where $q = \dfrac{G}{L_a + G}$
Identification:	Value of π assumed $= \pi^*$; $\qquad \alpha_1 = \beta_1/(1 + \beta_1)$
	$\phi = \beta_2/[\beta_1(\pi^* - 1) - 1]; \qquad \alpha_2 = \beta_3(1 - \alpha_1)/(\phi - 1)$
	$\alpha_0 = \beta_0/(1 + \beta_1) - \alpha_2; \qquad \alpha_i = \beta_{iq}(1 - \alpha_1)/(\phi - 1);$
	$\qquad\qquad\qquad\qquad\qquad\qquad\qquad i = S, Y, U, D$
	$\alpha_i{}^n = \alpha_i{}^a - \beta_i(1 - \alpha_1), \qquad i = S, Y, U, D$

effective prices and, simultaneously, the sensitivity of local allocatives to such price changes.

Model C—Federal Discrimination Among Grantees on the Basis of Grant Size

One interesting variation on the simple price-income equations of Model B is to relax the assumption that ϕ is constant across all grant-receiving communities. Let us ask whether the division of a grant between the fixed and variable segments in the multipart price varies systematically with the size of the grant itself. Do recipients of large grants face a proportionately higher or lower fixed cost of obtaining aid? Assuming ϕ is variable, specifically dependent on G, we write $\phi = d + eG$. Substituting into equation 18 gives a reduced form:

$$(21) \quad L_a = \frac{\alpha_0}{1 - \alpha_1} + \frac{\alpha_1}{1 - \alpha_1} L_n + \frac{(\alpha_1\pi - 1)}{1 - \alpha_1}(d + eG)G$$

$$+ \frac{\alpha_2}{1 - \alpha_1}[(d + eG - 1)q + 1] - \frac{\Sigma\alpha_i^n N_i}{1 - \alpha_1}$$

$$+ \frac{\Sigma\alpha_i^a N_i}{1 - \alpha_1}[(d + eG - 1)q + 1]$$

As in the case of the other models, this may be estimated with and without the auxiliary need variables S (population), Y (per capita income), U (percentage of urbanization), and D (population density). The structural parameters are identified as shown in Table 3. If numerical analysis shows e to be positive, then the monopoly power of the federal grantor will be shown to be less effective vis-à-vis local units receiving high per capita grants. If e is negative, the opposite conclusion obtains. In either case, such results would tend to confirm that the federal government effectively em-

TABLE 3. Identification of Model C

Structural equations: $L_a + \phi G = \alpha_o + \alpha_1(L_n + L_a + \pi\phi G) + \alpha_2\left(\dfrac{L_a + \phi G}{L_a + G}\right)$
$\qquad\qquad\qquad + (\Sigma\alpha_i{}^a N_i)\left(\dfrac{L_a + \phi G}{L_a + G}\right) - (\Sigma\alpha_i{}^n N_i)$

Reduced form: $\phi = d + eG$

$L_a = \beta_o + \beta_1 L_n + \beta_2 G + \beta_3 q + \beta_4 G^2 + \beta_5 Gq$
$\qquad + \beta_s S + \beta_y Y + \beta_u U + \beta_d D + \beta_{sq} Sq + \beta_{yq} Yq + \beta_{uq} Uq$
$\qquad + \beta_{dq} Dq + \beta_{sqg} SqG + \beta_{yqg} YqG + \beta_{uqg} UqG + \beta_{dqg} DqG$

Identification: $\alpha_1 = \beta_1/(1 + \beta_1); \quad \alpha_o = \dfrac{\beta_o}{1 + \beta_1} - \alpha_2; \quad \alpha = \beta_5\beta_2/(\beta_5\beta_2 - \beta_4\beta_3);$

$e = \beta_4\beta_5/(\beta_5\beta_2 - \beta_4\beta_3) \quad \alpha_2 = (\beta_5\beta_2 - \beta_4\beta_3)/\beta_4(1 + \beta_1)$

$\pi = [\beta_5\beta_2 - \beta_4\beta_3 + \beta_5(1 + \beta_1)]/\beta_5\beta_1$

ploys a monopolistic discriminatory power in the administration of the grant-in-aid system.

Models D and E—Tests for Identifying a Federal Supply of Grants Strategy

The foregoing price–income models of local allocation decisions might be extended in various ways. One obvious elaboration would involve tests of alternative functional forms. Local governments might be more or less sensitive to grant-induced price effects than the Stone–Geary form will allow. For example, a quadratic utility function might be explored. Rather than extending the basic model in those directions, however, we shall next examine the possibility that grants are distributed by the grantor in accord with some independent effort or equity-oriented criterion. In accord with this hypothesis, observed allocation outcomes are now assumed to be the result of three factors operating simultaneously, namely, the preferences of local governments, the price–income character of the grant, and the allocation preferences of the grantor.

To test for a federal grant strategy, one first could employ Model A as representative of the local allocation decision. Here, one tests the hypothesis that grants have only an ordinary income effect such that $0 < \pi < 1$. However, we now further suppose grants to be determined by federal law and administrative processes in accordance with the equity or efficiency rule: $G = b + cL_a$. This specification leads to reduced-form equations

$$(22) \quad L_n = \beta_0 + \beta_1 L_n + \beta_s S + \beta_y Y + \beta_u U + \beta_d D: \text{ and}$$
$$G = \theta_0 + \theta_1 L_n + \theta_s S + \theta_y Y + \theta_u U + \theta_d D$$

Table 4 shows the parameter identification results for this model. The model is underidentified; given an assumed value for $\pi = \pi^*$, however, one can derive the remaining structural parameters.

An important feature of this analysis is that a federal grant strategy, if discernible in the data, may appear to reward local effort rather than to compensate for local need. In this case, a pure income effect plus a federal

TABLE 4. Identification of Model D

Structural equations: $L_a + G = \alpha_o + \alpha_2 + \alpha_1 L_a + \alpha_1 L_n + \alpha_1 \pi G$
$\qquad + (\alpha_s^a - \alpha_s^n)S + (\alpha_y^a - \alpha_y^n)Y + (\alpha_u^a - \alpha_u^n)U + (\alpha_d^a - \alpha_d^n)D$
$\qquad G = b + cL_a$

Reduced form: $\qquad L_a = \beta_0 + \beta_1 L_n + \beta_s S + \beta_y Y + \beta_u U + \beta_d D$
$\qquad G = \theta_0 + \theta_1 L_n + \theta_s S + \theta_y Y + \theta_u U + \theta_d D$

Identification: $\qquad$ Value of π assumed $= \pi^*$; $\qquad \alpha_1 = (\beta_1 + \theta_1)/(1 + \beta_1 + \theta_1 \pi^*)$

$$C = \theta_1/\beta_1; \qquad b = \left[\frac{\beta_0}{\beta_1} - \frac{\theta_0}{\theta_1}\right] \bigg/ \left[\frac{\alpha_1 \pi^* - 1}{\alpha_1} - \frac{1 - \alpha_1}{\alpha_1 C}\right]$$

$$\alpha_2 + \alpha_0 = \frac{\beta_0}{\beta_1}\alpha_1 - b(\alpha_1 \pi^* - 1)$$

TABLE 5. Identification of Model E

Structural equations: $L_a + \phi G = \alpha_0 + \alpha_2 + \alpha_1(L_n + L_a) + \alpha_1\pi\phi G + \alpha_2(\phi - 1)q$
$$+ \Sigma(\alpha_i{}^a - \alpha_i{}^n)N_i + \Sigma\alpha_i{}^a(\phi - 1)N_iq$$
$$G = h + kL_a$$

Reduced form: $L_a = \beta_o + \beta_1 L_n + \beta_2 q + \beta_s S + \beta_y Y + \beta_u U$
$$+ \beta_d D + \beta_{sq}Sq + \beta_{yq}Yq + \beta_{uq}Uq + \beta_{dq}Dq$$
$$G = \theta_o + \theta_1 L_n + \theta_2 q + \theta_s S + \theta_y Y + \theta_u U$$
$$+ \theta_d D + \theta_{sq}Sq + \theta_{yq}Yq + \theta_{uq}Uq + \theta_{dq}Dq$$

Identification: $h = \left[\theta_o - \beta_o \dfrac{\theta_1}{\beta_1}\right];$ $k = \dfrac{\theta_1}{\beta_1}$ or $\dfrac{\theta_2}{\beta_2}$

effort-rewarding strategy would provide an alternative explanation for observed federal–state–local allocations and thus rival the price hypotheses developed in Model B.

An alternative to either of these models taken separately, however, is a combination of the two. Accordingly, the last model to be considered posits two structural equations

(23) $\quad L_a + \phi G = \alpha_o + \alpha_2 + \alpha_1(L_a + L_n) + \alpha_1\pi\phi G + \alpha_2(\phi - 1)q$
$$+ \Sigma(\alpha_i^a - \alpha_i^n)N_i + \Sigma\alpha_i^a(\phi - 1)N_iq$$

(24) $\qquad\qquad\qquad G = h + kL_n$

Equation 23 describes the local allocation decision or demand for aided functions (as dependent on both price and income effects of grants), while equation 24 describes the federal allocation strategy, or the supply of grants. Two reduced-form equations are derived by eliminating one and then the other dependent variable from equations 23 and 24. Identification of parameters h and k is shown in Table 5.

Conclusions, Qualifications, and Extensions of the Analysis

This chapter has developed one research tool for the analysis of economic transactions among governments. It is the first, to my knowledge, which attempts to decompose price and income effects of grants, in a way which will allow the two components to be estimated statistically.

Due to a variety of limitations, the analysis reported in this paper must necessarily be preliminary. The first limitation stems from the fact that we purposely have ignored prior information on the operation of grant systems. In a more detailed operational analysis of particular programs, detailed a priori knowledge will doubtless be available to the analyst. Obviously such information should not be discarded.

A second problem relates to the aggregation of data implicit in the models. Disaggregation is required in two dimensions. First, the appropriate unit to represent local government decision makers is needed. In our formulas, many transactions between state and local governments and many among localities within a state have been netted out in the data. Depending upon the grant program analyzed, a more appropriate definition of the local unit could be cities or metropolitan areas. In this case, new models would be required to differentiate between state and federal aid.

A second type of disaggregation required is along functional lines, between programs. Local responses to aid for education, welfare, highways, and so on, may differ greatly, and no one would claim that federal incentives are uniform across these diverse programs. This type of disaggregation, however, will confront the analyst with another problem. Specifically, it will require simultaneous estimation of local reactions to multiple functional grants. As an example, consider applying our price–income model to two functional categories. It is clear in principle that equations for both categories must be simultaneously estimated, because price and income effects from every grant influence expenditures on all aided categories. This obstacle, while difficult, may not be insurmountable, however. Appropriate information developed in a more detailed study of the institutions through which grants are channeled may allow us to generate reliable estimates of state–local response parameters, disaggregated by program or function. For example, knowledge of grants processes may lead the investigator to fix values of the parameter ϕ for some programs; cross-price effects may be known or assumed negligible; or, again, constraints on the fungibility of federal money may be summarized through interview and sampling. In other words, one way or another, detailed study of institutions should allow restrictions to be placed on relevant coefficients. Procedurally one might set up the full, disaggregated multiple sector, econometric system, see where problems of identification or irreducible bias arise, and then set about a study of the practices and procedures in various grant programs— a study designed specifically to circumvent the bias or identification problems previously discerned.

And last, within the context of disaggregation by function, a more extensive study should treat capital and current expenditures separately, and should distinguish between local expenditures financed out of local tax, and those financed out of local borrowing.

6 "Automatic" Increases in Tax Revenues—The Effect on the Size of the Public Budget

This chapter addresses what has become a commonplace assertion concerning the fiscal difficulties of state and local governments: the inability of these governments to meet their rapidly expanding budgetary needs because of the relatively low income elasticity of their revenue systems. The argument typically runs as follows. Because of rising relative costs of most public services and the growing demand for these services as incomes expand, the budgets of state and local governments necessary to keep pace with the demand for public outputs increase more than proportionately with aggregate personal income. Most state and local tax structures, however, possess an income elasticity not much in excess of unity, so that revenue "needs" grow more rapidly than do actual revenues at existing tax rates. This shortfall is sometimes described as a "revenue gap," implying that political obstacles to raising tax rates and introducing new sources of revenues are likely to result in an underprovision of public services.

* Professor of Economics, Princeton University.
    I am very grateful to William Baumol, Ray Fair, Stephen Goldfeld, James Ohls, Richard Quandt, and Michelle White for their comments on an earlier draft of this paper. In addition, I want to acknowledge the invaluable assistance of Judith Hawkes and John Murray with the empirical work and the expert help of Rosemary Little in locating the necessary data. For support of this study, I am indebted to the National Science Foundation.

The proposed solution to the problem is to make available to state and local governments more income-elastic sources of funding to close, at least partially, these revenue gaps. Such proposals have figured as central arguments, first, in the case for revenue sharing, and, second, in movements for reform of state and local tax systems themselves. Early in the push for revenue sharing, Walter Heller contended that, "At the Federal level, economic growth and a powerful tax system, interacting under modern fiscal management, generate new revenues faster than they generate new demands on the Federal purse. But at the state-local level, the situation is reversed. Under the whiplash of prosperity, responsibilities are outstripping revenues."[1]

Similarly, in the literature on state–local tax reform, one encounters with some regularity conclusions like those expressed by the New Jersey Tax Policy Committee, a body appointed by former Governor William Cahill to assess the New Jersey tax system and to make recommendations.

> The inelastic nature of New Jersey's state tax system means inevitable recurrent fiscal crises as revenue gaps open up. The expedients of the past can no longer be relied upon to close these gaps. Exploiting the same tax sources cannot cope with the projected growth in expenditures and provide the funds required to reform the tax structure. . . .
>
> If the "mix" of the tax system were changed to rely more heavily on income-elastic taxes, this would minimize the need for raising rates. The total tax burdens would be unchanged, but the need for recurrent legislative intervention would be diminished.[2]

I do not want to overstate the case. Obviously, revenue sharing and tax reform have had other fundamental objectives, such as a more equitable distribution of the tax burden. Nevertheless, I think it is fair to say that the goal of increasing the income elasticity of state–local sources of revenues has figured importantly in these programs.

One author, James Buchanan, has attributed an almost cosmic importance to the issue. Buchanan argues that, "with the adoption of the Sixteenth Amendment to the Constitution, the central government was granted access to the single fiscal weapon that was to remake the whole national fiscal pattern."[3] For, as Buchanan has indicated elsewhere, "In a period of rapidly increasing national product, that tax institution charac-

---

[1] See Heller, *New Dimensions of Political Economy* (Cambridge, Mass.: Harvard University Press, 1966), p. 118.

[2] *Summary of the Report of the New Jersey Tax Policy Committee* (Trenton: Feb. 23, 1972), pp. 2–3.

[3] See Buchanan, "Financing a Viable Federalism," in Harry L. Johnson, ed., *State and Local Tax Problems* (Knoxville: University of Tennessee Press, 1969), p. 5.

terized by the highest elasticity will tend, other things equal, to generate the largest volume of public spending."[4]

This last statement is a precise formulation of the hypothesis that I want to investigate here, and I stress that it is a hypothesis. It has served as an unexamined assumption in the sorts of discussions cited earlier. However, it is an empirical proposition, and, although it possesses a certain pragmatic plausibility, I will suggest in the next section that its validity, on a priori grounds, is far from clear. In fact, on one interpretation, it can be seen to imply some very curious and suspect behavior on the part of individual taxpayers. The later sections of the chapter consist of a set of empirical studies of state and local finances designed to answer the following question, Over the decade 1960-70 (a period of intense fiscal pressure on state and local governments), did those governments with more income-elastic revenue systems exhibit a comparatively large expansion in expenditures? I will also explore some international data bearing on this thesis as it relates to the growth of the public sector as a whole.

On the Income Elasticity of the Revenue System

First, I want to explore somewhat more systematically the proposition that the growth in public expenditure depends on the income elasticity of the revenue system. In terms of rational economic behavior, this seems a very peculiar assertion. We would expect that individual demands for public services would be based upon tastes, levels of income, and the cost of these services. With a given cost-sharing scheme (or set of tax shares), we could then determine the pattern of individual demands at some moment in time.

But why should people care about the income elasticity of the tax structure? What the proposition under study seems to imply is that people will not object to increases in public expenditure if they can be funded with no increases in tax rates (that is, from increments to revenues resulting solely from growth in income), but they will not support an expanded public budget if it requires a rise in tax rates. This suggests what people care about is not their tax *bill*, but rather their tax *rate*. Viewed this way, the hypothesis simply is not consistent with our conventional description of rational behavior; it implies that consumer–taxpayers are subject to a kind of "fiscal illusion."

[4] See Buchanan, *Public Finance in Democratic Process* (Chapel Hill: University of North Carolina Press, 1967), p. 65. Michael Reagan appears to share Buchanan's view: "Not least among the reasons for federal dominance of the revenue picture is the superior *elasticity* of the income tax, which gives it a considerable political advantage. . . ." See Reagan, *The New Federalism* (New York: Oxford University Press, 1972), p. 38.

A simple view of individual rationality, therefore, does not support the basic proposition. If people want a higher level of public services, they should presumably be willing to support the extended budget whether it is financed by an expansion in the tax base at given rates or by higher rates applied to a static base. The tax bill is what matters to the rational taxpayer. It is worth noting here that, over the decade of the 1960s, the remarkable expansion in state local public spending was financed largely from new levies consisting both of higher tax rates and the imposition of new forms of taxes. As Walter Heller and Joseph Pechman have pointed out, "Between 1959 and 1967, every state but one raised rates or adopted a major new tax; there were 230 rate increases and 19 new tax adoptions in this period."[5] On the face of it, there seem to be good reasons to be skeptical of the importance of an elastic tax structure for the level of public expenditure.

What kind of a case can we muster in support of the proposition? Somewhat surprisingly, hardly anyone[6] has really bothered to justify the hypothesis; it is typically put forward simply as an assertion of unquestioned validity. One encounters a few vague, unsatisfactory statements like that of L. L. Ecker Racz, "The people can afford higher taxes if only they would agree to the need. Instead they have an instinctive aversion to taxes for reasons none of us can be sure about. Public officials, interested in political longevity, in electoral support, feel obliged to heed the people's voice and echo their complaints."[7]

To try to make some sense of the hypothesis, we can go either (or, perhaps, both) of two routes. The first simply argues that this is a matter on which consumer–taxpayers do in fact act irrationally. There is a "faulty perception" on the part of the individual who regards an extra dollar of taxes as more costly to him if it results from a higher tax rate than from a more elastic base. At least over some range of tax liabilities, he suffers from a very direct sort of fiscal illusion. Richard Wagner, for example, advances this explanation:

> There also exists a large body of casual evidence supporting the existence of faulty fiscal perception. Whenever a legislative assembly considers changes in

[5] See Heller and Pechman, "Questions and Answers on Revenue Sharing," in *Revenue Sharing and Its Alternatives: What Future for Fiscal Federalism?*, Hearings before the Subcommittee on Fiscal Policy of the Joint Economic Committee (1967), pp. 11–17; reprinted in *Studies in Government Finance Reprints*, Brookings Institution Reprint No. 135, p. 10.

[6] The one important exception, to my knowledge, is Richard Wagner.

[7] See Ecker-Racz, *The Politics and Economics of State–Local Finance* (Englewood Cliffs, N.J.: Prentice-Hall, 1970), p. 197. This comment is in the context of a discussion of the growth of fiscal "needs" and revenues, and the elasticity of the tax structure.

tax rates, tortuous discussion takes place and considerable publicity is given to the deliberations. Yet no similar agonizing takes place over the continual, automatic increase in tax rates that is produced by progressivity in the national tax structure. Everyone is aware of a consciously enacted tax surcharge; a similar surcharge is enacted each year when income grows under progressive taxation, but many taxpayers remain unconscious of this surcharge.[8]

This is not, I think, an explanation to be dismissed lightly. There are other empirical studies which provide support for the presence of "illusory" behavior. William Branson and Alvin Klevorick, for example, have found the evidence on consumption behavior to be consistent with a model incorporating a money illusion; their findings indicate that in the short run, consumption expenditure depends not only on real income but also on a distributed-lag term involving the level of prices.[9]

Nevertheless, although this line of argument possesses a certain plausibility, I (along with most economists, I suspect) am uncomfortable with a hypothesis founded solely on irrational behavior. It is tempting to take an alternative tack to see if there are not some elements in the system of collective choice (that is, political mechanism) which can account for the supposed phenomenon. Perhaps the process through which individual tastes become translated into public budgets contains a set of incentives or costs that can explain the expansionary implications of a more income-elastic revenue system. In this case, the so-called fiscal illusion may be a property of the political mechanism and may yet be consistent with rational individual behavior.

One possible explanation can be found in the *transactions costs* associated with changes in tax rates. Richard Wagner has also made this case:

As viewed by legislators, the cost of thus changing tax structures may exceed the benefits, in which case the required change in relative tax collections will not occur. Such legislative behavior may be interpreted as a form of habitual behavior, and may very well be rational. There is a cost incurred in failing to routinize many types of activity—the value of the resources consumed in examining the consequences of the alternative choices, in this case, choices among possible tax structures. If the costs incurred by continual reexamination exceed the benefits received from taking more timely actions, habitual behavior is efficient—it is the least inefficient of the attainable alternatives.[10]

---

[8] See Wagner, *The Fiscal Organization of American Federalism* (Chicago: Markham, 1971), p. 87.

[9] Branson and Klevorick, "Money Illusion and the Aggregate Consumption Function," *American Economic Review*, vol. 59 (December 1969), pp. 832–849).

[10] Wagner, "The Fiscal Organization," pp. 85–86.

The New Jersey Tax Policy Committee may have had a similar point in mind when they argued that a more income-elastic tax structure would diminish "the need for recurrent legislative intervention."[11]

While there may be some truth to this contention, I have real reservations as to its likely importance. The resource costs of legislative action are, I should guess, relatively small compared to the benefits of meeting the service demands on the public sector. (In fact, there may be, on net, benefits from a regular reevaluation of spending *and* tax rates, like that, for example, which takes place with the annual budget referendum in many school districts; such processes require a periodic reconsideration of the level and composition of the public budget in the explicit context of the level of taxes.) I should be surprised if the costs of legislative action were sufficient to prevent significant realignments of the budget deemed desirable by the public. What may be true, as Wagner suggests, is that the need for adjustment in tax rates introduces some delays into the process. Thus, although I doubt that over the longer haul such transactions costs are likely to have much effect on the budget, there may be some short-run effects. The interesting question here, as for many other economic issues, may be, How long is the short run? More attention will be paid to this question in the empirical work which follows.

Aside from legislative transactions costs, the income elasticity of the tax structure may influence the budget because of taxpayer ignorance (*not* irrationality). As Anthony Downs has argued, it may be fully rational for the taxpayer–voter to be uninformed about the public budget.[12] Since the probability that an individual's vote will influence the outcome of a given state or local election is for most purposes negligible, it may not be in his interest to absorb the "costs" of becoming informed, or even of voting for that matter. "Ignorance of politics is not a result of unpatriotic apathy; rather it is a highly rational response to the facts of political life in a large democracy."[13]

The fiscal illusion (or "faulty fiscal perception" as Wagner called it earlier) need not, therefore, imply irrational behavior by taxpayer–voters. Because of the information costs inherent in determining the implicit "surcharge" resulting from rising income, the individual may not bother to make himself aware of the rise in his tax bill. In contrast, he can hardly help noticing the political turmoil and publicity that normally accompany a legislative action to increase tax rates. The information costs are, in this

---

[11] *Summary of the Report*, p. 3.

[12] Downs, "An Economic Theory of Political Action in a Democracy," *Journal of Political Economy*, vol. 65 (April 1957), pp. 135–150.

[13] Ibid., p. 147.

instance, very low; the daily perusal of the newspaper or other news media will call the increase to his attention. From the perspective of the politician, this fact suggests that he can, in effect, hide the costs of a larger budget from the taxpayer if he can finance it by an implicit surcharge from a growing tax base; if he must call attention to the higher taxes by a legislated increase in tax rates, he will have to incur some loss of taxpayer–voter support.

Again, this line of argument possesses a certain plausibility. I would be reluctant to push it too hard, however, for the logical implication is that if the tax system were sufficiently income-elastic, the public sector over time could come to absorb virtually all the taxpayer's income with no opposition. Pushed to its extreme, the argument is absurd. Nevertheless, it may yet be true that *over some range of tax bills* the fiscal illusion is operative.

With this as background, we turn in the next three sections to a series of empirical tests of the hypothesis that a higher income elasticity of the tax structure provides a positive stimulus to the growth of public spending. In the next two sections, we will examine the relationship between the change in expenditure for state governments and for a sample of local governments, respectively, and the elasticity of the tax structure over the period 1960–70. In a later section, I provide some additional, but admittedly highly conjectural, evidence using international cross-sectional data.

Growth in Spending by State Governments, 1960–70

The decade 1960–70 was one of extraordinary increases in state–local budgets; in the aggregate, state–local spending grew from a level of $52 billion in 1960 to $131 billion in 1970 (an average annual rate of increase of about 9 percent as compared with an average annual increase of roughly 6.5 percent in money GNP). The state and local public sector was indeed one of the economy's leading "growth industries" during the sixties. If a relatively inelastic tax structure acts as a constraint on state–local spending, this should be a period during which this constraint would be readily evident.

In this section, I will examine the growth in spending by state governments from 1960 to 1970 to see if, other things equal, those state governments with more income-elastic tax systems did in fact experience comparatively large increases in expenditure. The method of analysis is as follows. First, I postulate an expenditure function for state governments of the form:

(1) $$G_{it} = \alpha_0 + \alpha_1 V_{it} + \alpha_2 R_{it} + \alpha_3 S_{it} + u_{it},$$

where

G_{it} = per capita general expenditure of the ith state government in year t

V_{it} = socioeconomic determinants of public spending (a vector of socioeconomic variables) for state i in year t

R_{it} = federal grants per capita to the ith state government in year t

S_{it} = percentage of state–local spending in state i during year t that is undertaken by the state government

u_{it} = the disturbance term.

The rationale for most of these variables is pretty obvious. The variable S_{it}, for example, serves as a measure of the division of fiscal functions between the state and the local governments in each state. A relatively large value of S indicates a comparatively major role for the state government which, other things equal, should be reflected in a higher level of expenditure per capita.

Our concern, however, is with the growth in spending from 1960 to 1970. This we can express by simply subtracting the expenditure function for 1970 from that for 1960 to obtain:

(2) $$\Delta G_i = G_{i1970} - G_{i1960}$$
$$= \alpha_1(V_{i1970} - V_{i1960}) + \alpha_2(R_{i1970} - R_{i1960})$$
$$+ \alpha_3(S_{i1970} - S_{i1960}) + (u_{i1970} - u_{i1960}).$$

The hypothesis is that, in addition to the determinants of the growth in spending specified in equation 2, the income elasticity of the tax structure also influences the extent of the increase. To test this, I will add a measure of the tax elasticity to equation 2. We then proceed to estimate the equation[14]:

(3) $$\Delta G_i = a_0 + \alpha_1(V_{i1970} - V_{i1960}) + \alpha_2(R_{i1970} - R_{i1960})$$
$$+ \alpha_3(S_{i1970} - S_{i1960}) + \alpha_4 T_i + u_i^*,$$

where

T_i = a measure of the income elasticity of the tax structure of the state government in state i

u_i^* = the associated disturbance term.

A central concern is to determine a reliable operational measure of the elasticity of the state tax system. There is one set of estimates available

---

[14] While the formulation of the tax-elasticity hypothesis in equation 3 seems plausible, one would prefer a specification founded securely on a more compelling conceptual framework. For an exploration of this issue, see Appendix B.

from an external source. The Advisory Commission on Intergovernmental Relations (ACIR), in a report published in 1968, has provided an estimate for each state of the income elasticity of the state government's tax structure for 1967.[15] There are, however, a number of deficiencies in the ACIR estimates. For example, the estimates are based on only a fraction of the sources of each state's tax revenues; for some states, this is less than 60 percent. For purposes of this study, however, my most serious reservation is that the estimates are for a single year. What we need are measures of the tax elasticity for the decade 1960–70; where the ACIR variable is used in the equations that follow, we must, therefore, treat it as a proxy for tax elasticity for the decade. As such, it does not reflect either the timing of *changes* in the tax structure that occurred prior to 1967 or *any* effects of changes instituted after 1967. Incidentally, one interesting feature of the ACIR estimates is the considerable range in income elasticity that they indicate; the estimates extend from 0.7 for Nebraska to 1.4 for Oregon. This is encouraging, for it suggests substantial variation in the crucial independent variable.

To obtain more reliable measures of tax elasticity, I assembled fiscal histories of the forty-eight coterminous states over the decade, 1960–70. Existing estimates of the income elasticities of various state taxes indicate that far and away the most income-elastic is the individual income tax. Most estimates are in the range of 1.5 to 2.0; the ACIR suggests a "medium" estimate of 1.65. This compares to estimates of the income elasticity of general sales taxes of roughly unity. State corporate income taxes also exhibit a somewhat above-average income elasticity; the ACIR indicates a "medium" value of 1.2.[16]

This suggests that the extent of reliance on income taxation should provide a reasonable approximation to the relative elasticity of the tax structure. For this purpose, the comparative importance of the individual income tax should be the better measure. As noted above, state individual income taxes are, on average, much more income-elastic than corporation income taxes. Moreover, the individual income tax is typically a much larger component of state tax systems than are corporate income taxes: over the decade 1960–70, state governments, in the aggregate, collected 14 percent of their tax revenues from individual income taxes and only

[15] Advisory Commission on Intergovernmental Relations (ACIR), *Sources of Increased State Tax Collections: Economic Growth Vs. Political Choice*, Information Report M-41 (October 1968), table 7.

[16] For a summary of estimated income elasticities for various state and local taxes, see ACIR, *Federal–State Local Finances: Significant Features of Fiscal Federalism* (February 1974), table 173, p. 320; for the "medium" estimates see ACIR, *Sources of Increased State Tax Collections*, p. 3.

5.5 percent from taxes on corporation income. There is, incidentally, considerable variation in the degree of reliance on income taxation: there are a handful of states (Florida, Nevada, Ohio, Texas, Washington, and Wyoming) which generated no tax revenues from either individual or corporate income taxation during 1960–70, while at the other extreme the state of Oregon derived 47 percent of its tax revenues from the individual income tax alone and 57 percent from individual and corporation income taxes combined over this same period.

In the regression equations that follow, I will employ in succession the following different measures of the income elasticity of the state government tax system for 1960–70:

T_{Ai} = ACIR estimate of the income elasticity of the tax structure of ith state government for 1967

T_{Ii} = the sum of individual income tax receipts over the years 1960–70 as a percentage of the sum of total tax receipts over the same years for the ith state government

T_{Ci} = the sum of corporation income tax receipts (1960–70) as a percentage of total tax receipts (1960–70) for state i

$T_{Ti} = T_{Ii} + T_{Ci}$ = total income tax receipts as a percentage of total tax receipts for 1960–70 for state i.

I used each of these four tax variables in the process of estimating equation 3. The estimation also required making explicit the socioeconomic determinants of state government expenditure. I have used the following three variables:

Y_{it} = median family income in state i during year t

L_{it} = percentage of families in state i with incomes below the poverty line in year t

P_{it} = population size of state i in year t.

Table 1 summarizes the results obtained from estimating equation 3 by ordinary least squares (OLSQ). In Table 1, equation 1.1 employs the ACIR tax variable; although its sign is consistent with the hypothesized effect of tax elasticity, the standard error is sufficiently large that we cannot reject the null hypothesis of no effect at a .05 level of significance. In contrast, the individual income tax variable, T_I, is statistically significant in equation 1.2 of Table 1; this result does support the hypothesized positive effect of tax elasticity on the growth in public expenditure. As expected, the individual income tax variable appears to possess a more reliable effect on the growth of public spending than do corporation tax receipts.

There is good reason, however, to be suspicious of the results in Table 1, because there are some obvious systems problems inherent in the inter-

TABLE 1. Growth in State Government Expenditures, 1960–70 (OLSQ)

Equation	Constant	$(Y_{1970} - Y_{1960})$	$(L_{1970} - L_{1960})$	$(P_{1970} - P_{1960})$	$(R_{1970} - R_{1960})$	$(S_{1970} - S_{1960})$	T_A	T_I	T_C	T_T	R^2
(1.1)	−84 (1.3)	43 (3.4)	1.1 (0.9)	−16 (2.1)	2.1 (6.8)	−0.8 (0.7)	43 (1.5)				.64
(1.2)	−29 (0.5)	37 (3.1)	1.0 (0.8)	−15 (2.1)	2.0 (7.1)	−0.8 (0.7)		1.1 (2.9)			.69
(1.3)	−12 (0.2)	35 (2.5)	1.5 (1.2)	−18 (2.3)	2.0 (6.5)	−0.8 (0.7)			2.4 (1.9)		.65
(1.4)	−16 (0.3)	34 (2.8)	1.2 (1.0)	−15 (2.1)	2.0 (6.9)	−0.8 (0.7)				0.9 (3.0)	.69
(1.5)	−15 (0.2)	34 (2.6)	1.2 (1.0)	−15 (2.1)	2.0 (6.7)	−0.8 (0.7)		0.9 (2.3)	1.0 (0.8)		.69

Notes:

Dependent variable: ΔG = change in per capita expenditure by the state government, 1960–70.

$N = 48$ for all equations.

The numbers in parentheses below the estimated coefficients are the absolute values of the t statistic.

relationships among the variables. Two of the independent variables are clearly endogenous to the system as a whole. Intergovernmental grants are in part determined by the level of public expenditure, since under matching-grant programs, the funds received depend upon the level of spending selected by the recipient. The state share variable will also obviously depend upon the level of state government expenditure. This means that we must regard the variables $(R_{1970} - R_{1960})$ and $(S_{1970} - S_{1960})$ as endogenous to the complete system of equations; OLSQ can thus be expected to generate biased estimators of the coefficients in equation 3.

To provide a more reliable set of parameter estimates, I estimated the same set of equations by two-stage least squares; Table 2 indicates these results.[17] The general pattern of findings is quite similar to those obtained with OLSQ. In particular, the elasticity variable, T_I, which indicates the extent of reliance on individual income taxation, is significant at a .05 level under both methods of estimation. In both cases, the estimated coefficient is about unity. This suggests what I would judge to be a nontrivial, yet modest, effect on the growth in spending. As an illustration, the point estimates indicate that a state government which generated 35 percent of its revenues through individual income taxes would, other things equal, have experienced an expansion in spending per capita over 1960–70 of roughly $35 per capita more than a state which collected no revenues from this source. But this compares to a mean increase in state expenditure per capita of $228 for the decade. I think this can hardly be regarded as a really "large" effect.

The estimated coefficient for the ACIR tax variable, although positive in sign, remains statistically insignificant at a .05 level. Similar remarks also apply to the estimated coefficient of the tax elasticity variable defined in terms of corporation tax revenues. In sum, the evidence from state government fiscal experience in the decade 1960–70 does appear consistent with the basic hypothesis: state governments with more income-elastic tax structures did, other things equal, increase expenditure per capita by a greater amount than those with less elastic tax systems. The estimated magnitude of the effect does, however, appear to be of a modest order.

[17] To "purge" the endogenous grant and share variables of their presumed correlation with the disturbance term, I employed in the first stage a set of instrumental variables including, in addition to other independent variables in equations 1.1 through 1.5, the following: the percentage of the population under age eighteen, the percentage of the population over age sixty-five, the percentage of elementary and secondary pupils enrolled in nonpublic schools, the percentage of population that is black, acres of federally owned land per capita, miles of highway per capita, population density, and percentage of the population living in urban areas. These I regard as exogenous variables which would influence (directly or indirectly) the level of grants received and/or the state's share of state–local expenditures.

TABLE 2. Growth in State Government Expenditures, 1960–70 (TSLS)

Equation	Constant	$(Y_{1970} - Y_{1960})$	$(L_{1970} - L_{1960})$	$(P_{1970} - P_{1960})$	$(R_{1970} - R_{1960})$	$(S_{1970} - S_{1960})$	T_A	T_I	T_C	T_T
(2.1)	-145 (1.3)	54 (2.7)	0.5 (0.3)	-19 (2.1)	2.5 (3.6)	-1.2 (0.3)	34 (1.0)			
(2.2)	-102 (0.9)	49 (2.5)	0.4 (0.2)	-18 (2.1)	2.5 (3.8)	-1.2 (0.3)		1.0 (2.4)		
(2.3)	-63 (0.5)	44 (1.9)	1.4 (0.7)	-20 (2.3)	2.4 (3.2)	-2.2 (0.6)			2.0 (1.3)	
(2.4)	-77 (0.7)	44 (2.2)	0.9 (0.5)	-18 (2.2)	2.4 (3.6)	-1.7 (0.5)				0.9 (2.4)
(2.5)	-76 (0.6)	44 (2.0)	0.8 (0.4)	-17 (2.1)	2.4 (3.4)	-1.4 (0.4)		0.9 (2.1)	0.6 (0.4)	

Note: Dependent variable: ΔG = change in per capita expenditure by the state government, 1960–70. The numbers in parentheses below the estimated coefficients are to absolute values of the t statistic.

A few other findings from the regression equations in Tables 1 and 2 are noteworthy in passing. First, the parameter estimates suggest that intergovernmental grants have a large stimulative effect: for each extra dollar of grants received from the federal government, state governments exhibit an increase in spending of roughly $2.00 to $2.50. Second, the cross-sectional results point to an income elasticity of spending well in excess of unity: a state with a medium family income which rose $1,000 more than that of another state experienced, on average, a per capita increase in state government expenditure of about $40 to $50 more than the latter. Evaluated at the mean values of these variables, this implies an income elasticity of spending of approximately 1.5. Finally, it is interesting to note that population growth shows a consistent and significant negative association with the expansion in the budget. States apparently have not increased per capita spending at a pace commensurate with population growth.

Growth in Spending by City Governments, 1960–70

In addition to studying state governmental finances, I examined the fiscal experiences of a sample of thirty-three U.S. cities over this same decade of 1960–70. The well-documented fiscal difficulties of the cities during this period suggest at first glance that the constraints of an inelastic tax structure (if indeed such constraints exist) should have been at least as effective on city, as on state, budgetary growth.

To discern these effects, I ran essentially the same series of tests on increases in city expenditures as those reported for the states in the preceding section. My initial expectation was that these effects, even if operative, would be more difficult to uncover for the cities. First, cities typically place less reliance on income taxes than do the states (although there are a few striking exceptions); and second, to the extent that they do rely on income taxes, the taxes are normally a good deal less progressive than those of the states.[18] For both these reasons, the extent of reliance on income taxes probably does not reflect such substantial differences in tax elasticity for the cities as for the states. Moreover, during this decade many of the cities

---

[18] Most city income taxes are simply a flat percentage (for example, 0.5 or 1 percent) of taxable income with some level of exemption as the only source of progression. There are some exceptions; New York City, for example, possesses a progressive rate structure that in 1968 ranged from 0.4 percent on taxable income less than $1,000 to 2 percent on taxable income in excess of $30,000. State income taxes, in contrast, typically exhibit substantial progression. For a comprehensive description of the structure of individual state and city income taxes in 1968, see Advisory Commission on Intergovernmental Relations, *State and Local Finances, Significant Features 1966 to 1969* (November 1968), tables 35–42.

experienced a net outflow of population, including many middle-income and upper-income families, so that the expansion in the income tax base (even in per capita terms) was considerably less than at the state level.

The sample of thirty-three cities is, incidentally, not a random sample.[19] It draws on the larger cities in the United States and includes in particular all those sizable cities which made a significant use of income taxes; this was necessary to obtain a substantial variation in the tax variable. I should also mention that the Census of Governments does not provide any division between receipts from individual and corporate income taxes for the cities. The tax variable for the cities is simply total income taxes as a percentage of total tax revenues. Within the sample, this variable had a mean of 11.7 percent and ranged in value from 70 percent for Columbus, Ohio, to zero for twenty-one of the cities.

The regression results using OLSQ are reported in equation 4, where ΔX is the change in variable X from 1960–70; the variables Y, L, P, and R are defined as earlier except that they now refer to city jurisdictions; S is again the state percentage of total state–local public expenditure, and T is income tax receipts as a percentage of total tax revenues.

$$(4) \qquad \Delta G = 121 - .03\Delta Y - 1.3\Delta L - .02\Delta P + 1.5\Delta R$$
$$\qquad\qquad (1.5) \quad (1.3) \qquad (0.4) \qquad (0.3) \qquad (12.6)$$
$$\qquad\qquad + 5.6\Delta S + 1.1T \qquad\qquad\qquad R^2 = .89$$
$$\qquad\qquad (2.5) \quad (2.7) \qquad\qquad\qquad\qquad N = 33$$

As for the states, the estimated equation is consistent with the hypothesis of a positive effect of tax elasticity on the growth in public spending. However, ΔR and ΔS are again clearly endogenous variables. Estimating the equation by TSLS yields[20]:

$$(5) \quad \Delta G = 125 - .03\Delta Y - 1.2\Delta L - .02\Delta P + 1.5\Delta R + 6.2\Delta S + 1.1T .$$
$$\qquad\quad (1.4) \quad (1.3) \qquad (0.4) \qquad (0.2) \qquad (8.8) \qquad (1.7) \qquad (2.7)$$

The results in equation 5 are quite similar to those in equation 4. In particular, the tax variable is still positive and significant at a .05 level. Curiously, the magnitude of the estimated coefficient is virtually identical with that for the states—about unity. This indicates that for each extra

[19] The list of cities appears in Appendix A.

[20] The additional instrumental variables used to estimate equation 5 were percentage of population under age eighteen, percentage change of fraction of population under age eighteen, percentage of nonwhite population, percentage change of fraction of nonwhite population, miles of highway per capita in the state, population density in the state, population density in the city, and percentage of population in the state residing in urban areas.

percent of tax revenues raised through income taxes, our "typical" city over 1960-70 would have generated an extra $1.00 increase in public spending per capita. Since the mean increase in expenditure per capita for the sample of cities is $147, this suggests a *relatively* larger budgetary effect than at the state level. The point estimate indicates that a city which collected one-half of its revenues from income taxes would have experienced a budgetary increase of about $50 more than a city with no income taxation, a differential of approximately one-third of the mean increase in city spending.

Also, like the findings for the states, equations 4 and 5 suggest a strong expansionary impact of intergovernmental grants; an additional dollar of grant revenues to the cities is associated, on average, with an increase in expenditure of about $1.50. There are, however, some anomalies in the results. The change in medium family income, for example, is negatively related (although not statistically significant) to the growth in spending.

Tax-Elasticity and the Size of the Public Sector: Some International Evidence

The evidence presented in this section is not intended to be taken very seriously. My primary interest is simply to point out that there is, in principle, no reason to restrict the tax elasticity hypothesis to state and local governments. As Buchanan's earlier comments suggest, it is certainly conceivable that the growth of the public sector as a whole may depend to some extent on the structure of the tax system.

Since I had an extensive body of international cross-sectional data readily available from an earlier study, I undertook some admittedly crude statistical work which bears on this proposition. I stress that these data did not permit the estimation of equations like those in the preceding sections, where the growth in public expenditure was regressed on the *changes* in the set of explanatory variables. Instead, I will present findings based on *levels* of variables, where the equation is of the general form:

$$(6) \qquad W_i = \alpha_0 + \alpha_1 V_i + \alpha_2 T_i + u_i,$$

where

W_i = total government revenues as a percentage of national income in country i

V_i = vector of socioeconomic variables in country i influencing the relative size of the public sector

T_i = measure of the income elasticity of country i's tax structure

u_i = disturbance term.

Equation 6 says that the relative *size* of the public sector depends on a set of variables, V, and a measure of tax elasticity, T. The presumption here is that T serves as a proxy for the historic level of the income elasticity of the country's tax system; if this elasticity has been comparatively high, then we would expect, other things equal, to find the country with a relatively large public sector.

Other studies have identified certain systematic influences on the relative size of the government sector.[21] The most powerful explanatory variable (as embodied in Wagner's "law") is the level of per capita income; wealthier societies typically possess a comparatively large "fisc." In addition, the degree of openness seems to have a systematic positive association with the governmental share of economic activity. For V, I will therefore employ measures of per capita income and the size of the foreign trade sector. A good measure of tax elasticity is hard to come by. We can, however, approximate the types of measures used in earlier sections with a variable provided in the International Bank for Reconstruction and Development's *World Tables:* direct taxes as a percentage of government general revenue. Direct taxes "comprise all taxes and surtaxes levied as a charge on the income of households and private nonprofit institutions (including contributions to social security); corporate income and excess profits taxes; and taxes on undistributed profits or on capital stock which are levied at regular intervals.[22] The tax variable can thus be roughly described as income taxes as a percentage of government revenues, and we would expect this to be positively related to the income elasticity of the tax system as a whole.

For a sample of fifty-seven countries,[23] the estimated equation is:

$$(7) \qquad W = 3.8 + .007Y + .15F + 0.3T \qquad R^2 = .64$$
$$\qquad\qquad (1.3) \quad (3.6) \quad\; (2.1) \quad (3.3)$$

where:

Y = gross domestic product (GDP) per capita in U.S. dollars for 1965
F = total exports as a percentage of GDP for 1965
W and T are defined as noted above (also for 1965).

---

[21] For a survey of these findings, see Kilman Shin, "International Difference in Tax Ratio," *Review of Economics and Statistics*, vol. 51 (May 1969), pp. 213–220.

[22] International Bank for Reconstruction and Development *World Table 6* (Washington, D.C.: IBRD, 1968), p. 1. The *World Tables* are a rich source of data assembled and distributed periodically by the research staff of the Economic Program Department of the IBRD.

[23] The sample consists of the fifty-eight countries listed in the data appendix to Wallace E. Oates, *Fiscal Federalism* (New York: Harcourt Brace Jovanovich, 1972), with the exception of the Malagasy Republic for which the data were incomplete.

The tax elasticity variable is highly significant: we can reject the null hypothesis of no association at a .01 level. In addition, the magnitude of the effect suggested by the point estimate of the coefficient is quite sizable; other things equal, an increase of 10 percentage points in the percentage of revenues raised by direct taxation is associated with a rise of 3 percentage points in tax revenues as a percentage of national income.

The absence of any very compelling case for the particular specification of the equation is admittedly unsettling. Nevertheless, the tax elasticity variable does possess considerable explanatory power. The evidence, crude as it is, from this international data is consistent with the findings for U.S. state and local governments: tax elasticity does appear to be positively related to the growth (or, in this case, the relative size) of the public budget.

Some Reflections and Conclusions

In concluding this study, I want to make four observations. First, there is one further possible source of bias in the results. For state and city governments, we found that over the decade 1960-70 the income elasticity of the tax structure did indeed display a positive association with the growth in public expenditure. This is certainly consistent with the basic hypothesis under study. Yet it is also conceivable that those states desiring a more rapid expansion in public outputs could have consciously constructed a more income-elastic revenue system to finance the anticipated budgetary growth. Or, in other words, one could argue that there is an element of taste reflected in the tax structure itself; more specifically the contention could be that, not only does tax elasticity influence the size of the budget, but planned increases in the budget also affect the income elasticity of the tax system. In fact, if one simply treats the tax-elasticity variable as endogenous in the TSLS estimations, the variable does in fact lose much of its explanatory power. This possibility suggests that the findings in this paper must be accepted with some caution.

Second, it is important to remember that the primary empirical support provided by this study comes from a single decade of rapid expansion in state and city budgets. The findings suggest that in the short run (in this case, ten years), a relatively inelastic tax structure may slow somewhat a process of very rapid growth in the public budget. It may yet be the case that over the longer haul the income elasticity of the tax system exerts little influence on the growth of the public sector. This latter issue remains unclear, although the international cross-sectional findings provide some tentative support for a long-run effect as well.

Third, Michelle White has called to my attention the limitations which result from confining the study of U.S. jurisdictions to states and large cities. In a letter to me, she suggests the hypothesis that ". . . there is more reluctance to raising tax rates at the state level and in large cities than in smaller governmental units such as school districts and townships. This may reflect the fact that increases in statewide taxes cause increased income redistribution so the middle class pay for more than they get. This is less true of local taxes."

Finally, I want to emphasize that this chapter is directed to the *empirical* question of whether or not tax elasticity influences the size of the public budget. I have deliberately avoided the issue of the *desirability* of income-elastic revenue systems. This latter normative problem is a complex one. One may argue that a higher income elasticity of the tax structure is needed to allow the state–local public sector to overcome some type of fiscal illusion in the political process and thereby provide "adequate" levels of public services. This kind of argument would certainly find widespread support among elected officials and legislators; increased tax elasticity provides for them a greater flexibility in budgetary operations. But it is just this sort of flexibility that one may wish to constrain. In fact, a good case can be made for opting for a relatively inelastic structure that will force frequent recourse to the electorate. In this way, the taxpayer–voter will more often have the opportunity to register his support or opposition to public programs in full view of their cost. This latter matter is, of course, a meaningful one only if the hypothesized budgetary effects of tax elasticity are in fact true. The findings in this study would, on balance, lend support to the view that such effects do exist, although they may not be quantitatively very large.

Appendix A

TABLE A.1 Sample of Cities

Atlanta	Memphis
Baltimore	Milwaukee
Birmingham	Minneapolis
Boston	Newark, N.J.
Buffalo	New Orleans
Chicago	New York
Cincinnati	Norfolk
Cleveland	Oklahoma City
Columbus, Ohio	Omaha
Dallas	Philadelphia
Denver	Phoenix
Detroit	Pittsburgh
Fort Worth	Saint Louis
Houston	Saint Paul
Indianapolis	San Antonio
Kansas City, Mo.	Toledo
Louisville	

Appendix B

The equation that I chose to estimate in the text is of an admittedly *ad hoc*, pragmatic character. It is not formally derived from an underlying model of political choice, but is rather a reduced-form expenditure function of the sort frequently encountered in the empirical literature in state and local finance. Its most useful property is its simplicity for purposes of estimation.

Here, I would like to report on, and explore further, some difficulties encountered in an attempt to specify and estimate a somewhat more appealing description of budgetary dynamics. In particular, the equation I employed in the text postulates that the (per capita) dollar increase in public expenditure over the chosen time period depends on changes in the values of key socioeconomic and fiscal variables and on the income elasticity of the tax structure. One disturbing aspect of this formulation is that it makes no provision for any displacements from desired levels of spending or explicit lags in the adjustment process; the specification, for instance, allows for no impact on budgetary growth of the particular value of G in the initial year (except as reflected through its alleged determinants).

To correct this deficiency, I tried to conceptualize the process of budgetary change so as to account explicitly for a lagged adjustment to the desired level of public expenditure. Let us postulate that at any time, t, there is a desired level of spending, G_t^*, which depends on a vector of socioeconomic variables, V_t, and a stochastic term, u_t, such that:

$$(\text{B.1}) \qquad G_{it}^* = \gamma_0 + \gamma_1 V_{it} + u_{it},$$

for the ith jurisdiction at time t. However, the actual level of expenditure, G_t, depends not only on the desired level, but on the level of spending in the preceding period, $G_{(t-1)}$. There is thus a process of adjustment from the previous size of the budget to that desired in the present period.

A useful way to introduce the impact of the income elasticity of the tax system is to visualize it as influencing the speed of adjustment of the budgetary process. During a decade when the desired budget is growing quite rapidly, we might expect a highly elastic tax structure to facilitate the expansion of public expenditure. Legislative action to raise tax rates or to introduce new sources of revenues is itself time consuming and costly so that an inelastic tax system exerts a kind of drag on budgetary growth. This suggests a formulation like the following:

$$(\text{B.2}) \qquad G_{it} = f(T_i)[G_{it}^* - G_{i(t-1)}] + v_{it}.$$

Equation B.2 specifies that the fraction of the adjustment of the budget from its *actual* level in period $(t - 1)$ to its *desired* level at time t is a function of the income elasticity of the tax system plus the influence of the stochastic disturbance term, v_{it}.

The next step was the search for a specific functional form for equation B.2 that would make sense analytically and would permit estimation. A straightforward and reasonable specification is:

(B.3) $$G_{it} = [\beta_0 + \beta_1 T_i][G_{it}^* - G_{i(t-1)}] + v_{it}.$$

The test of the hypothesis thus centers on the sign and statistical significance of the coefficient, β_1. If the income elasticity of the tax system is of importance in the process of budgetary adjustment, we would expect β_1 to be positive and significantly different from zero. Otherwise, the degree of adjustment will be indicated by β_0 alone. The null hypothesis, therefore, is $H_0: \beta_1 = 0$.

It is impossible to estimate equation B.3 directly, because G_{it}^* is not observable. However, by substituting the expression in equation A.1 for G_{it}^* in equation B.3, we obtain an equation consisting of observed variables and disturbance terms. This is the equation I undertook to estimate[24]:

(B.4) $$G_{it} = [\beta_0 + \beta_1 T_i][\gamma_0 + \gamma_1 V_{it} - G_{i(t-1)}] + v_{it}.$$

The econometric obstacles to estimating equation B.4 are quite formidable. First, the equation is nonlinear both in the variables and parameters. And, second, recall from the text that some of the variables in the vector, V_{it}, are themselves dependent on G_{it}, which introduces a simultaneous-equation bias into the parameter estimators. The price of introducing a more satisfactory conceptual formulation has thus been the creation of a very difficult estimation problem.

There are techniques for estimating an equation like equation B.4, where there are endogenous variables and nonlinearities in both the parameters and variables (although the full range of properties of the estimators is not yet known).[25] One approach is a two-stage procedure in which the first stage involves "approximately purging" the endogenous independent variables of their correlation with the disturbance term by regressing them on polynomial forms of the predetermined variables in the system. The calculated values of the endogenous variables are then substituted for the

---

[24] One simplification was necessary immediately. The presence of both u_{it} and v_{it} (and various cross-product terms involving them) made the equation impossible to estimate by any technique I could discover. I, therefore, made the further assumption that equation B.1 is an exact relationship, and simply discarded u_t from equation B.1.

[25] Stephen M. Goldfeld and Richard E. Quandt, *Nonlinear Methods in Econometrics* (Amsterdam: North Holland Publishing Co., 1972), chap. 8; and Takeshi Amemiya, "The Nonlinear Two-Stage Least-Squares Estimator," Stanford University, Department of Economics, working paper no. 25 (June 1973).

observed values. In the second stage, the values of the parameters are determined so as to minimize the sum of the squared deviations of the predicted values of the dependent variable from their observed values.

Unfortunately, my attempts to estimate equation B.4 by this two-stage procedure yielded nonsensical results; the values of the estimated parameters were far outside the bounds of any reasonable expectations. This may reflect an inappropriate specification of the process of budgetary adjustment, or, alternatively, some inadequacies of the estimation technique. At any rate, I feel at this juncture that the results reported in the text represent the most solid evidence that I can offer on the tax-elasticity hypothesis.

For Product Safety Concerns and Information please contact our EU
representative GPSR@taylorandfrancis.com
Taylor & Francis Verlag GmbH, Kaufingerstraße 24, 80331 München, Germany